FOOD SCIENCE AND TECHNOLOGY

AUTHENTICATION OF VIRGIN OLIVE OIL USING NMR AND ISOTOPIC FINGERPRINTING

FOOD SCIENCE AND TECHNOLOGY

Additional books in this series can be found on Nova's website under the Series tab.

FOOD SCIENCE AND TECHNOLOGY

AUTHENTICATION OF VIRGIN OLIVE OIL USING NMR AND ISOTOPIC FINGERPRINTING

ROSA M. ALONSO-SALCES
JOSÉ M. MORENO-ROJAS
MARGARET V. HOLLAND
AND
CLAUDE GUILLOU

Nova Science Publishers, Inc.
New York

Copyright © 2011 by Nova Science Publishers, Inc.

All rights reserved. No part of this book may be reproduced, stored in a retrieval system or transmitted in any form or by any means: electronic, electrostatic, magnetic, tape, mechanical photocopying, recording or otherwise without the written permission of the Publisher.

For permission to use material from this book please contact us:
Telephone 631-231-7269; Fax 631-231-8175
Web Site: http://www.novapublishers.com

NOTICE TO THE READER

The Publisher has taken reasonable care in the preparation of this book, but makes no expressed or implied warranty of any kind and assumes no responsibility for any errors or omissions. No liability is assumed for incidental or consequential damages in connection with or arising out of information contained in this book. The Publisher shall not be liable for any special, consequential, or exemplary damages resulting, in whole or in part, from the readers' use of, or reliance upon, this material. Any parts of this book based on government reports are so indicated and copyright is claimed for those parts to the extent applicable to compilations of such works.

Independent verification should be sought for any data, advice or recommendations contained in this book. In addition, no responsibility is assumed by the publisher for any injury and/or damage to persons or property arising from any methods, products, instructions, ideas or otherwise contained in this publication.

This publication is designed to provide accurate and authoritative information with regard to the subject matter covered herein. It is sold with the clear understanding that the Publisher is not engaged in rendering legal or any other professional services. If legal or any other expert assistance is required, the services of a competent person should be sought. FROM A DECLARATION OF PARTICIPANTS JOINTLY ADOPTED BY A COMMITTEE OF THE AMERICAN BAR ASSOCIATION AND A COMMITTEE OF PUBLISHERS.

Additional color graphics may be available in the e-book version of this book.

LIBRARY OF CONGRESS CATALOGING-IN-PUBLICATION DATA
Authentication of virgin olive oil using NMR and isotopic fingerprinting /
authors: M. Rosa Alonso-Salces ... [et al.].
   p. cm.
 Includes index.
 ISBN 978-1-61122-309-5 (softcover)
 1. Olive oil--Analysis. 2. Olive--Mediterranean Region--Identification.
3. Nuclear magnetic resonance. 4. Mass spectrometry. I. Alonso-Salces, M. Rosa.
 TP683.A98 2010   664'.3620287--dc22   2010038627

Published by Nova Science Publishers, Inc. † New York

Contents

Preface		vii
Abbreviations		ix
Chapter 1	Introduction	1
Chapter 2	Materials and Methods	5
Chapter 3	Results and Discussion	11
Acknowledgment		49
References		51
Index		57

PREFACE

In this study three harvests of virgin olive oils (VOOs) from the Mediterranean basin were analyzed by ^{1}H NMR and ^{13}C and ^{2}H isotope ratio mass spectrometry (IRMS) to determine their geographical origin at the national, regional, or PDO level. An increase in ^{13}C/^{12}C and ^{2}H/^{1}H ratios was observed in VOOs from northern to southern regions; VOOs produced in France and northern Italy presented higher values than those produced in Spain and Greece. This trend was also observed in VOOs from northern and southern Italy, and for those from southern Spain and the other Spanish regions. Moreover, significant differences were found between years of harvest due to climatic conditions (rainfall, temperature) and, in some cases, between samples from the same region which could be due to such factors as orography, distance from the sea, or a particular microclimate. ^{1}H NMR fingerprints of VOOs were analyzed by principal component analysis, linear discriminant analysis (LDA), and partial least-square discriminant analysis (PLS-DA). LDA and PLS-DA achieved consistent results for the characterization of PDO *Riviera Ligure* VOOs. PLS-DA afforded the best model: for the Liguria class, 92% of the oils were correctly classified in the modeling step, and 88% of the oils were properly predicted in the external validation; for the non-Liguria class, 90 and 86% of hits were obtained, respectively. A stable and robust PLS-DA model was obtained to authenticate VOOs from Sicily: the recognition abilities were 98% for Sicilian oils and 89% for non-Sicilian ones, and the prediction abilities were 93 and 86%, respectively. More than 85% of the oils of both categories were well predicted in the external validation. Greek and non-Greek VOOs were properly classified by PLS DA: >90% of the samples were correctly predicted in the crossvalidation and external validation. Stable isotope data, that is, $\delta^{13}C$ and δ^2H, provided complementary geographical information to the ^1H NMR fingerprints of the VOOs.

ABBREVIATIONS

VOO	virgin and extra virgin olive oils;
PDO	Protected Designation of Origin;
PGI	Protected Geographical Indication;
TSG	Traditional Specialty Guaranteed;
NMR	nuclear magnetic resonance;
FID	free induction decays;
NIR	near-infrared;
MIR	middle-infrared;
IR	infrared;
FT	Fourier transformation;
GC	gas chromatography;
LC	liquid chromatography;
SPE	solid-phase extraction;
ANOVA	analysis of variance;
PCA	principal component analysis;
PC	principal component;
LDA	linear discriminant analysis;
PLS-DA	partial least-squares discriminant analysis;
PRESS	predicted error sum of squares;
RMSEP	root-mean-square error of prediction;
CV	crossvalidation;
LOO	leave-one-out crossvalidation.

Chapter 1

INTRODUCTION

1.1. PRODUCTION AND COMMERCIALIZATION OF VIRGIN OLIVE OIL IN THE EUROPEAN UNION

Olive oil is the oil extracted exclusively from the fruit of *Olea europea* L. only by means of mechanical methods or other physical procedures that do not cause any alteration of the glyceric structure of the oil and preserve its characteristic properties (*1*). The International Olive Oil Council (IOOC) establishes the definitions and classes of olive oils, based on methods of production and the free acidity of the oil, as well as the rules for their commercialization (*2*). Virgin and extra virgin olive oils (VOO) are the only ones that can be commercialized on the international markets (*2*). At present, 77% of the global production of olive oil takes place in the Mediterranean basin, mainly in Spain, Italy and Greece. The characterization of the geographical origin of VOO is becoming increasingly important. VOOs are permitted to be marketed under a Protected Designation of Origin (PDO), Protected Geographical Indication (PGI), or Traditional Specialty Guaranteed (TSG) label, on the basis of their area and methods of production [Council Regulations (EEC) No 2081/92 and No 2082/92]. According to the EU definition, PDO products are most closely linked to the concept of *terroir* — a sense of place discernible in the flavor of the food. PDO products must be produced, processed and prepared in a specific region using traditional production methods. The raw materials must also be from the defined area whose name the product bears. The quality or characteristics of the product must be due essentially or exclusively to its place of origin, that is,, climate, the nature of the soil and local know-how. Food products with a PGI status must have a geographical link in at least one of the stages of production, processing or preparation. The European Commission has already registered in the "Register of protected designations of origin and protected geographical indications" 95 PDO and PGI olive oils, produced in Spain, Italy, Greece, Portugal, France, and Slovenia. As can be expected, given the financial benefits associated with these prestigious labels, it is very likely that economic fraud occurs (e.g., labeling a non-PDO product as a PDO one or adulteration with olive oils that do not fulfill the PDO requirements). Other fraudulent practices that were detected by the state security forces were the adulteration of olive oils with low-grade oils and the mislabeling of olive oils. For instance, olive oil imported into Italy from Tunisia, Greece, and Spain was relabeled as the finest Italian product. Other ploys include labeling inferior quality oil as extra virgin olive oil and claiming European Union (EU) subsidies for growing olives in Italy while actually importing them from elsewhere. Currently EU laws allow the commercialization of foreign olive oil as belonging to an EU country if it is cut with a small amount of the domestic product. However, the EU is about to establish new labeling rules that will make origin labeling compulsory for virgin and extra virgin labeled olive oil. The rules propose that oil produced from olives from just one EU country will have to be labeled with the name of the country of origin. Oil produced from olives from more than one EU country would be labeled as a 'blend of community olive oils', while oil produced using olives from outside the EU would be labeled as a 'blend of non-community olive oils' or 'blend of community and non-community olive oils'. Therefore, analytical methods are urgently needed to guarantee the authenticity and traceability of PDO and PGI olive oils, as well as the country of provenance of the olive oil, to help prevent illicit practices in this sector, and to support the antifraud authorities dealing with these issues.

1.2. CHARACTERIZATION OF VIRGIN OLIVE OILS

More than 98% of VOO is made up of triglycerides and the remaining 1–2% of minor components such as squalene, α-tocopherol, phytosterols, phenolic compounds, carotenoids, and aliphatic and terpenic alcohols, which constitute the unsaponifiable fraction of the oil (*1*). The chemical composition of this fraction may vary both qualitatively and quantitatively depending on vegetal species, climatic conditions, extraction and refining procedures and storage conditions, which also greatly influence the organoleptic quality and stability of the oil (*1*). The diversity and interdependence among all of these factors makes it highly unlikely that these influences would be the same in different regions. Therefore, the geographical characterization of VOO addresses all these agronomic, pedoclimatic, and botanical aspects that are unique to the oil of each origin (*3*).

A considerable number of sensorial (*4*), physical (*5*), and chemical (*5, 6*) approaches combined with statistical analysis have been used to distinguish olive oils from different types, botanical, geographical origins, and pedoclimatic conditions. For this purpose, fatty acids (*7*), triglycerides (*8*), sterols (*9*), phenolic compounds (*10*), pigments (*11*) have been analyzed by conventional methods that usually required time consuming pre-treatment methods (solvent extraction, isolation and/or derivatization) followed by chromatographic techniques (*12*) such as GC-MS and/or GC-FID (*13, 14*), and HPLC-MS (*10*). PDO olive oils were distinguished using physicochemical parameters of the oils and chemometric class-modeling tools (*5*), sensory parameters and fatty acid profiles of the oils (*7*), or the oil sterol composition (*9*).

Fingerprinting techniques such as NMR (*15, 16*), NIR (*17*), MIR (*18*), FT-IR, FT-MIR, and FT-Raman (*19-22*) spectroscopies, MS (*23*), GC×GC-Tof-MS (*24-26*), and DNA fingerprinting (*27, 28*) have been used for the determination of food authenticity (*29*). These types of techniques are particularly attractive because they are nonselective, require little or no sample pretreatment; use small amounts of organic solvents or reagents, and typically require only a few minutes per sample. Chemometric analysis of NIR spectra of VOOs allows us to determine its composition and geographical origin (*30*). ^{1}H, ^{13}C, and/or ^{31}P NMR analysis of the bulk oil (*31, 32*) or the unsaponifiable fraction of olive oil (*33*), in combination with multivariate techniques, have been used to distinguish VOOs according to their geographical origin. ^{1}H NMR and the more recently developed hyphenated LC-SPE-NMR technique have been applied to study phenolic compounds in the polar fraction of olive oil for authentication purposes (*34*).

The stable isotope ratio of $^{13}C/^{12}C$ gives a straightforward information about the primary photosynthetic metabolism of plant products (*35*), and $^{18}O/^{16}O$ and $^2H/^1H$ ratios are good indicators of environmental conditions (*36*). The study of $\delta^{13}C$ variability of olive oils in several harvests showed that it was not dependent on either the degree of ripeness or maturity state of the olives or the olive variety (*37*). The isotopic fractionation of C and H is linked to pedoclimatic factors (soil, climate, and latitude); therefore, these data may contribute to the geographical discrimination of olive oils. Thus, isotope ratio mass spectrometry (IRMS) methods have also been used for the geographical characterization of olive oils. $^{13}C/^{12}C$ and $^{18}O/^{16}O$ ratio measurements on olive oil evidenced that the samples clustered according to the latitude and climate of the production areas (*38*).The distance from the sea and environmental conditions (water stress, atmospheric moisture and temperature) during the plant growth were found to be important factors that contributed to the variability of the oils. $^{18}O/^{16}O$ measurements on Spanish olive oils showed that their geographical origin was the main factor responsible for the variability of the samples (*39*). $^{13}C/^{12}C$ and $^{2}H/^{1}H$ ratios of Italian VOOs distinguished olive oils produced at different latitudes (*40*). $^{13}C/^{12}C$, $^{18}O/^{16}O$, and $^{2}H/^{1}H$ measurements were performed on VOOs from eight European production sites to discriminate olive oils according to their geographical origin (*41*).

In the present study, a statistically significant number of authentic VOOs from seven countries, namely, Italy, Spain, Greece, France, Turkey, Cyprus, and Syria, from three different harvests (2004/05, 2005/06, and 2006/2007) were characterized by ^{1}H NMR and ^{13}C and ^{2}H IRMS with authentication purposes. The ^{1}H NMR fingerprints were analyzed by pattern recognition and classification techniques, such as principal component analysis (PCA), linear discriminant analysis (LDA), and partial least-square discriminant analysis (PLS-DA), to evaluate the best approach to identify the geographical origin at the national, regional, and/or PDO level. Further isotopic measurements of $\delta^{13}C$ and $\delta^{2}H$ were performed on the samples by IRMS to help with the geographical discrimination of VOOs. This work was developed within the framework of the EU TRACE project (http://www.trace.eu.org) with the aim of supporting antifraud authorities in dealing with the prevention and detection of illicit practices in the olive oil sector. Moreover, this study is also of interest to consumer, honest oil producer, and regulatory bodies , since it will contribute to ensure the authenticity and traceability of such a high value foodstuff.

Chapter 2

MATERIALS AND METHODS

2.1. CHEMICALS AND PLANT MATERIAL

Deuterated chloroform for NMR analysis (99.8 atom % D) was provided by Sigma-Aldrich Chemie (Steinheim, Germany).

Virgin olive oils (963 samples) from seven countries of the Mediterranean basin, namely Italy (661 VOOs), Spain (144 VOOs), Greece (97 VOOs), France (39 VOOs), Turkey (14 VOOs), Cyprus (6 VOOs), and Syria (2 VOOs), were collected directly from the producers (olive oil mills) from most of the main producing regions of these countries during three harvests (2004/05, 2005/06, and 2006/2007). The sample collection was carried out with the collaboration of Laboratorio Arbitral Agroalimentario (Ministry of Agriculture and Fishery, Spain), General Chemical State Laboratory D'xy Athinon (Greece), General State Laboratory (Ministry of Health, Cyprus), Departamento de Química Orgánica - Universidad de Córdoba (Spain), Istituto di Metodologie Chimiche (CNR, UNAPROL, Dipartimento di Chimica e Technologie Farmaceutiche ed Alimentari, Università di Genova, Italia), Fondazione Edmund Mach (Istituto San Michele all'Adige, Italy), and Eurofins Scientific Analytics (France), within the framework of the EU TRACE project. The true type (virgin or extra virgin) and origin of the olive oils at the national, regional, and PDO level were assured.

VOOs produced in the Mediterranean basin are usually defined as multivarietal because of the presence of several olive cultivars in the same field; from 3 or 4 different varieties to as many as 70, depending on the PDO or production area. Sampling for the present study was planned so as to cover the maximum variability related to the harvests, olive varieties, and production areas. The Italian samples were representative of the olive oil producing areas, which are markedly influenced by pedoclimatic factors which vary considerably from the north to the south of the country.

2.2. NMR ANALYSIS

Aliquots of 40 µL of each VOO were dissolved in 200 µL of deuterated chloroform, shaken in a vortex, and placed in a 2 mm NMR capillary. The ^{1}H NMR experiments were performed at 300 K on a Bruker (Rheinstetten, Germany) Avance 500 (nominal frequency 500.13 MHz) equipped with a 2.5 mm broadband inverse probe. The spectra were recorded using a 7.5 µs pulse (90°), an acquisition time of 3.0 s (32K data points), a total recycling time of 4.0 s, a spectral width of 5500 Hz (11 ppm), and 64 scans (+ 4 dummy scans), with no sample rotation. Prior to Fourier transformation, the free induction decays (FIDs) were zero-filled to 64K, and a 0.3 Hz linebroadening factor was applied. The chemical shifts are expressed in δ scale (ppm), referenced to the residual signal of chloroform (7.26 ppm) (*42*). The spectra were phase- and baseline-corrected manually. The multivariate data analysis was performed on a region of the NMR spectra between 0 and 7 ppm. The spectra were binned with 0.02 ppm wide buckets and normalized to total intensity over the region 4.10–4.26 ppm (glycerol signal). TopSpin 1.3 (2005) and Amix-Viewer 3.7.7 (2006) from Bruker BioSpin GMBH (Rheinstetten, Germany) were used to perform the processing of the spectra. The data table generated with the spectra of all samples was then used for pattern recognition. Eight buckets in the region 4.10–4.26 ppm (reference region) were excluded in the multivariate data analysis.

2.3. IRMS ANALYSIS

Isotopic measurements of $δ^{13}C$ were performed by continuous flow IRMS using a Carlo Erba elemental analyzer (EA) EA-1108-CHN (Thermo Fisher, Milan, Italy) coupled to a DeltaPlus mass spectrometer (Thermo Fisher, Rodano, Italy). The $δ^{13}C$ signal for the reference peak was 4000 mV; the oxidation column temperature, 1050 °C; the reduction column temperature, 650 °C; and the GC column temperature, 65 °C. $δ^2H$ measurements were carried out by continuous flow IRMS using a total conversion elemental analyzer (TC/EA) coupled to a Delta PlusXP mass spectrometer (ThermoFisher, Rodano, Italy). The δ^2H signal for the reference peak was 7000 mV; the GC column temperature, 80 °C; and the glassy-carbon column reactor temperature, 1450 °C. Aliquots of 0.3 mg of VOO were weighed in tin capsules for determination of $\delta^{13}C$; and silver capsules, for δ^2H.

The results of the carbon ($\delta^{13}C$) and hydrogen (δ^2H) isotope ratio analyses are reported in per mile (‰) on the relative δ scale and refer to the international standards V-PDB (Vienna Pee Dee Belemnite) for the carbon isotope ratio and V-SMOW (Vienna Standard Mean Ocean Water) for the hydrogen isotope ratio. All results were calculated according to the equation:

$$\delta\,(‰) = [(R_{Sample}/R_{Reference}) - 1] \times 1000$$

where R is the ratio of the heavy to light stable isotope (e.g. $^2H/^1H$) in the sample (R_{Sample}) and in the standard ($R_{Reference}$). The calibration of the control gases (CO_2 and H_2) was performed using the following reference materials: (*i*) for $\delta^{13}C$ measurements, IAEA-CH7-Polyethylene ($\delta^{13}C$ = −32.15%) and IAEACH6-Sucrose ($\delta^{13}C$ = −10.4%) for CO_2 gas cylinder calibration; and (*ii*) for δ^2H measurements, IAEA-CH7-Polyethylene (δ^2H = −100.3%) and V-SMOW (δ^2H = 0%) for H_2 gas cylinder calibration. An olive oil sample was calibrated with the international reference materials previously mentioned and used as a working standard. This standard was analyzed at regular intervals to control the acceptable repeatability of the measurements, and to correct for any possible drifts. The standard deviation (n = 10) determined using the corresponding reference gas were 0.05% for $\delta^{13}C$, and 0.8% for δ^2H. Each olive oil sample was analyzed in triplicate, the standard deviations being <0.15% for $\delta^{13}C$, and <2.7% for δ^2H.

2.4. DATA ANALYSIS

The data set, made up of the values of the 342 buckets of the ^1H NMR spectra (variables in columns) measured on the 963 VOOs analyzed (samples in rows), was firstly analyzed by univariate procedures (ANOVA, Fisher index and box–whisker plots), and afterwards, by the following multivariate techniques which are already described in the literature cited (*43*): unsupervised ones as principal component analysis (PCA); and supervised ones as linear discriminant analysis (LDA), and partial least-squares discriminant analysis (PLS-DA). Statistical and chemometric data analysis were performed by means of the statistical software packages Statistica 6.1 (StatSoft Inc., Tulsa, OK, USA, 1984–2004), The Unscrambler

9.1 (Camo Process AS, Oslo, Norway, 1986–2004), and SIMCA-P 11.0 (Umetrics AB, Umea, Sweden, 1992–2005).

In LDA, the variable selection strategy was the following. First, a modified best subset selection was used, which is a variable selection procedure that performs a search for the best subsets of a small number of variables that fulfill the criterion for choosing the best one (Wilks' lambda, rate of misclassification, etc.). This can be computed relatively quickly and in several steps: first, best subset selection is applied to the complete data matrix to obtain the first best (small) subset of variables; then, in a second step, best subset selection is sued on a data set omitting the variables selected in the first step, a second best subset is achieved, and so on. Finally, a refined selection of the variables selected successively in the previous steps was performed using forward stepwise selection (*43*).

In PLS-DA, PRESS or RMSEP are plotted against the number of the principal components to find the optimal number of PLS components. Sometimes there are several almost equivalent local minima on the curve; the first one should be chosen to avoid overfitting (according to the principle of parsimony). The model with the smallest number of features should be accepted from among equivalent models on the training set. Once PLS components are estimated by crossvalidation, the classifications in the training-test set are represented in a box–whisker plot to define half of the distance between the quartiles as the boundary.

The supervised techniques were applied to the autoscaled (or standardised) or Pareto-scaled data matrix of the VOO profiles following these steps: (*i*) the data set was divided into a training-test set and an external data set; (*ii*) the training-test set was subsequently divided into a training set and a test set several times in order to perform crossvalidation; (*iii*) the training-test set was used for the optimization of parameters characteristic of each multivariate technique by crossvalidation, for instance the number of PLS components in PLS-DA or for variable selection in LDA; (*iv*) a final mathematical model was built using all the samples of the training-test set and the optimized parameters; (*v*) this model was validated using an independent test set of samples (external data set), that is,. performing an external validation. During the parameter optimization step, the models were validated by three-fold crossvalidation (3-fold CV) or leave-one-out crossvalidation (LOO). The reliability of the classification models achieved in the crossvalidation was studied in terms of recognition ability (percentage of the samples in the training set correctly classified during the modeling step) and prediction ability (percentage of the samples in the test set correctly classified by using the models developed in the training step). The reliability of the final model was evaluated in terms of classification ability (percentage of the samples of the training-test set correctly classified by using the optimized model) and the prediction ability in the external validation (percentage of the samples of the external set correctly classified by using the optimized model) (*43*).

Chapter 3

RESULTS AND DISCUSSION

3.1. ^{1}H NMR SPECTRA OF VIRGIN OLIVE OILS

^{1}H NMR spectra of the 941 VOOs produced in different PDO areas and/or regions of the EU olive oil producing countries, namely, Italy, Spain, Greece, and France, and 22 VOOs from other countries from the Mediterranean basin (Turkey, Cyprus, and Syria) were recorded. Olive oil is mainly made up of triglycerides, differing in their substitution patterns in terms of length, degree, and kind of unsaturation of the acyl groups, and by minor components such as mono- and di-glycerides, sterols, tocopherols, aliphatic alcohols, hydrocarbons, fatty acids, pigments, and phenolic compounds (*1*). The chemical shifts of the ^{1}H signals of the triglycerides are well-known (*16*). Minor oil components are only observed by ^{1}H NMR when their signals do not overlap with those of the main components and their concentrations are high enough to be detected (*33, 44-47*). **Figure 1** shows a typical ^{1}H NMR spectrum of a VOO, and **Table 1** gathers the common ^{1}H NMR signals of the major and some minor compounds together with their chemical shifts and their assignments to protons of the different functional groups (*33, 44-47*). Several signals of minor compounds were found in ^{1}H NMR of VOO since they were not overlapped by those of the triglyceryl protons: cycloartenol at 0.318 ppm and 0.543 ppm, β-sitosterol at 0.669 ppm, stigmasterol at 0.687 ppm, squalene at 1.662 ppm, sn-1,2 diglyceryl group protons at 3.71 ppm and 5.10 ppm, and three unknown terpenes at 4.571 ppm, 4.648 ppm and 4.699 ppm, as already observed by other authors (*33, 44-47*). The NMR spectra provided information about the most significant NMR signals or regions for the classification of VOOs according to their geographical origin.

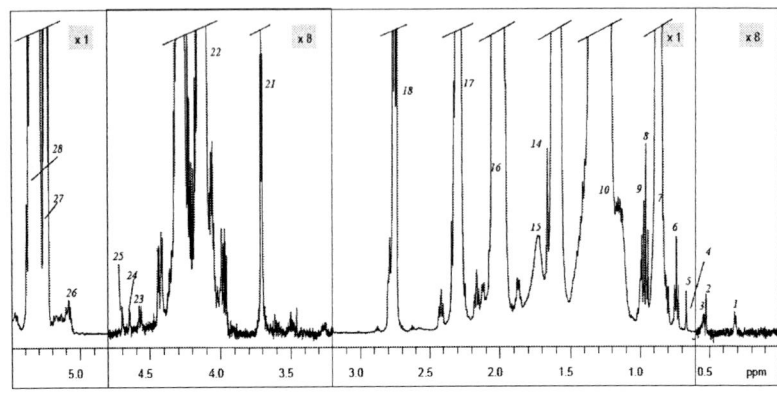

Figure 1. ^{1}H NMR spectra of a VOO (signal numbering, see Table 1).

3.2. INFLUENCE OF THE YEAR OF HARVEST ON ^{1}H NMR FINGERPRINT OF VIRGIN OLIVE OILS

The data set consisted of a 963 × 342 matrix, in which rows represented the 963 samples of VOO and columns the 342 buckets of the ^{1}H NMR spectrum. The presence of outliers in the dataset was analyzed by PCA, and 28 extreme samples from different origins and harvests were removed after the presence of some irregularities in their NMR spectra noted. The four first principal components (accounting for 63% of total system variability: PC1 for 31%, PC2 for 13%, PC3 for 11%, and PC4 for 7%) showed that samples were distributed into compact clusters, even though some subgroupings according to the year of harvest were observed. PC2, PC3, and PC4 contained information related to the year; however, **Figure 2** shows that the three clusters partially overlapped. Because 70% of the samples were Italian and the rest were from other countries in the Mediterranean region, seasonal aspects seem to affect all samples in the same way, independently of their geographical origin. Therefore, in the modeling for the authentication of agricultural food products, it is important to have chemical data from several harvests to obtain general classification models that include the seasonal variability, as well. On the other hand, the PCA score plots did not show any clusters related to the geographical origin or the PDO of the oils. This indicates that the direction of maximum variability in the data set did not correspond to the direction of maximum discrimination among the geographical origins or PDOs. This suggests the presence of other sources of variability. Indeed, the year of harvest was confirmed to be one of these as seen above.

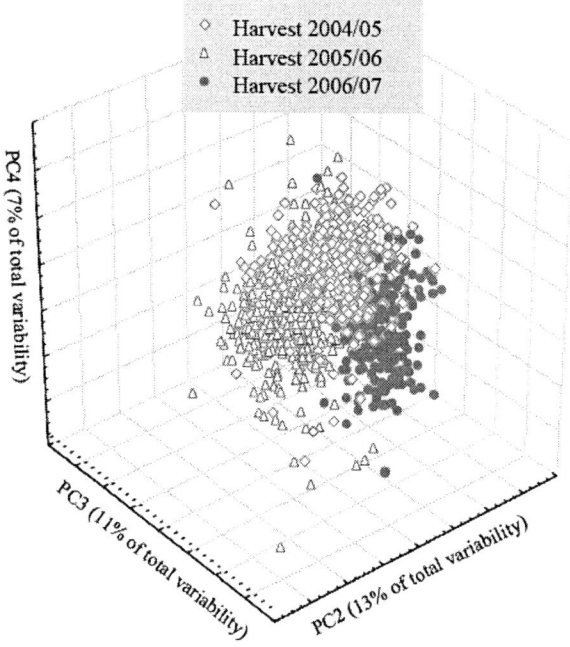

Figure 2. PCA score plot of the VOO samples on the space defined by PC2, PC3, and PC4.

3.3. δ^{13}C AND δ^{2}H OF VIRGIN OLIVE OILS BY IRMS

Camin et al. (40) demonstrated that δ^2H and δ^{18}O of bulk olive oils were correlated, therefore in the present study only δ^2H were determined in VOOs, as well as δ^{13}C. The δ^{13}C and δ^2H results were consistent with the expectations for plants following the C$_3$ cycle (48, 49). ^{13}C/^{12}C and ^{2}H/^{1}H ratios increased from northern regions to southern ones. Moreover, significant differences were found between the years of harvest due to climatic conditions (rainfall, temperature); and in some cases, also between samples from the same region, due to the influence of orography, distance from the sea, microclimate, etc. However, regarding olive varieties, no conclusive results were expected, since it is difficult to separate the influence of olive variety and pedoclimatic conditions on the isotopic signature of olive oils. This is due to the fact that usually, in each region, farmers cultivate the same varieties.

Table 1. Chemical shift assignments of the ^{1}H NMR signals of the main components of VOOs

#	Chemical shift (ppm)	Multiplicity[a]	Functional group	Attribution
1	0.318	d	-CH$_2$- (cyclopropanic ring)	cycloartenol
2	0.527	s	-CH$_2$- (cyclopropanic ring)	cycloartenol
3	0.543	d	-CH$_2$- (cyclopropanic ring)	cycloartenol
4	0.669	s	-CH$_3$ (C18-steroid group)	β-sitosterol
5	0.687	s	-CH$_3$ (C18-steroid group)	stigmasterol
6	0.740	t	-CH$_3$ (^{13}C satellite of signal at 0.87 ppm)	
7	0.866	t	-CH$_3$ (acyl group)	saturated, oleic (or ω-9) and linoleic (or ω-6)
8	0.960	t	-CH$_3$ (acyl group)	linolenic (or ω-3)
9	0.987	t	-CH$_3$ (^{13}C satellite of signal at 0.87 ppm)	
10	1.19-1.37		-(CH$_2$)$_n$- (acyl group)	
11	1.243		-(CH$_2$)$_n$- (acyl group)	saturated (palmitic, stearic)
12	1.256		-(CH$_2$)$_n$- (acyl group)	oleic
13	1.288		-(CH$_2$)$_n$- (acyl group)	linoleic and linolenic
14	1.51-1.65		-OCO-CH$_2$-CH$_2$- (acyl group)	
15	1.662	s	-CH$_3$	squalene
16	1.96-2.07		-CH$_2$-CH=CH- (acyl group)	
17	2.26-2.32	m	-OCO-CH$_2$- (acyl group)	
18	2.72-2.82		=CH-CH$_2$-CH= (acyl group)	
19	2.754	t	=CH-CH$_2$-CH= (acyl group)	linoleic
20	2.789	t	=CH-CH$_2$-CH= (acyl group)	linolenic
21	3.69-3.73	d	-CH$_2$OH (glyceryl group)	sn 1,2-diglycerides
22	4.09-4.32		-CH$_2$OCOR (glyceryl group)	triglycerides
23	4.571	d		terpene

#	Chemical shift (ppm)	Multiplicity[a]	Functional group	Attribution
24	4.648	s		terpene
25	4.699	s		terpene
26	5.05-5.15	m	>C*H*OCOR (glyceryl group)	*sn* 1,2-diglycerides
27	5.22-5.28	m	>C*H*OCOR (glyceryl group)	triglycerides
28	5.28-5.38	m	-C*H*=C*H*- (acyl group)	

[a] Signal multiplicity: s, singlet; d, doublet; t, triplet; m, multiplet.

3.3.1. Countries Producing Virgin Olive Oils

The median, minimum and maximum values of $\delta^{13}C$ and δ^2H for VOOs from Italy, Spain, Greece, and France are reported in **Table 2**. Spain and Greece showed higher values for both isotope ratios than Italy and France, which was in agreement with the results obtained by other authors (*37, 38*). The trend observed for the $\delta^{13}C$ and δ^2H values was Italy ≤ France < Greece ≅ Spain. Italian VOOs presented the largest variability of $\delta^{13}C$ and δ^2H values. These results can be explained by the fact that isotopic fractionation is highly influenced by climate conditions, which, in turn, are related to the distance from the sea and the latitude of the olive groves. In this sense, Italian samples also included a large variability due to olive cultivars.

Rainfall before harvesting also markedly affects the isotopic distribution of these elements: On the one hand, the lower values for δ^2H of the Italian VOOs from the harvests 2005/06 were due to the higher level of rainfall that took place in Italy during 2005 before harvesting (*50*).; on the other hand, the higher $\delta^{13}C$ values of the Greek VOOs from the harvest 2004/05 were influenced by the low level of rainfall which occurred six months before harvesting in Greece (*50*).

3.3.2. Italian Virgin Olive Oils

VOOs from the main olive oil producing regions of Italy were studied: Lombardia (LOMB), Trentino Alto Adige (TAA), Veneto (VEN), Liguria (LIG), Umbria (UMB), Lazio (LAZ), Abruzzo (ABR), Molise (MOL), Campania (CAMP), Puglia (PUG), Calabria (CAL), and Sicily (SIC) (Figure 3). The median, the minimum and the maximum values of $\delta^{13}C$ and δ^2H of the VOOs from each of these regions for the three harvests studied are plotted in box–whisker plots in Figure 4. The variability of the measurements was highly influenced by the harvest: $\delta^{13}C$ and δ^2H for the VOOs from harvest 2004/05 were higher in comparison to those from harvest 2005/06, and similar to those from harvest 2006/07. The enrichment in the heavier isotopomeres, ^{13}C and ^2H, (less negative values in δ-notation) is linked with higher temperatures and less rainfall. In fact, there was relatively heavy rainfall in 2005 (*50*), therefore VOOs harvested from October'05 to February '06 presented more negative isotopic values. $^{13}C/^{12}C$ and $^2H/^1H$ ratios increased from northern regions to southern ones for each harvest.

Table 2. Median, minimum, and maximum values of $\delta^{13}C$ and $\delta^{2}H$ of European VOOs from three harvests (2004/05, 2005/06, and 2006/07)

Harvest						2004/05			2005/06				2006/07			
Country	Parameter	N	Median	Min.	Max.	N	Median	Min.	Max.	N	Median	Min.	Max.
Italy	$\delta^{13}C$	226	-29.8	-31.7	-27.6	252	-30.1	-31.8	-28.5	184	-30.1	-32.1	-28.5
	$\delta^{2}H$		-145	-160	-129		-150	-170	-139		-148	-170	-137
Spain	$\delta^{13}C$	59	-29.3	-30.2	-28.1	38	-28.4	-30.0	-27.5	34	-28.7	-30.1	-28.1
	$\delta^{2}H$		-142	-150	-129		-140	-154	-125		-139	-150	-132
Greece	$\delta^{13}C$	43	-28.6	-29.7	-26.5	46	-29.2	-30.5	-28.3	7	-29.1	-29.8	-28.4
	$\delta^{2}H$		-142	-149	-130		-142	-164	-135		-146	-150	-143
France	$\delta^{13}C$	9	-29.5	-30.0	-28.8	10	-30.0	-30.3	-29.4	20	-29.4	-30.5	-28.4
	$\delta^{2}H$		-147	-157	-140		-151	-157	-140		-151	-163	-138

Figure 3. Italian political regions.

VOOs from northern regions, that is, Lombardia, Trentino Alto Adige, and Veneto presented the lowest $\delta^{13}C$ values for the three harvests. In contrast, the VOOs from the southern regions, that is, Sicily and Calabria, showed the highest values of $\delta^{13}C$ for the harvests studied. Campania and Puglia, which are southern regions situated in the north of Calabria (**Figure 3**), produced VOOs that presented $\delta^{13}C$ values close to those for VOOs from regions in the centre of Italy, except for the VOOs from Campania of the harvest 2004/05. $\delta^{13}C$ values for VOOs are likely to be positively related to the vicinity to the sea and dryness of the climate but negatively related to latitude (*38, 51*). This trend was already observed in wines (*52*). All of these climatic parameters influence, in different ways, the availability of water, relative humidity, and temperature, which in turn control the stomatal aperture and the internal CO_2 concentration in the leaf (*35*). This means that the $\delta^{13}C$ varies from enriched to depleted values depending on the previously mentioned geographical and climatic conditions.

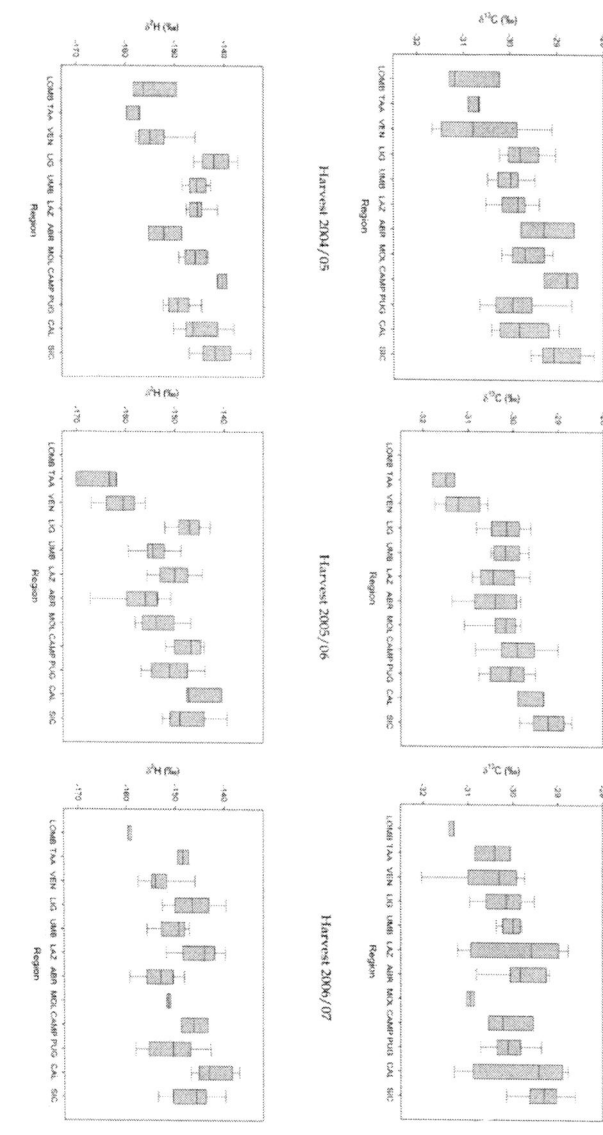

Figure 4. $\delta^{13}C$ and δ^2H values of Italian VOOs. Number of samples from each region in harvests 2004/05, 2005/06, and 2006/07, respectively: Lombardia (LOMB) 5, 1, and 3 VOOs; Trentino Alto Adige (TAA) 3, 3, and 3 VOOs; Veneto (VEN) 10, 14, and 10 VOOs; Liguria (LIG) 63, 79, and 68 VOOs; Umbria (UMB) 18, 20, and 9 VOOs; Lazio (LAZ) 29, 17, and 13 VOOs; Abruzzo (ABR) 6, 9, and 9 VOOs; Molise (MOL) 13, 17, and 3 VOOs; Campania (CAMP) 7, 14, and 8 VOOs; Puglia (PUG) 28, 30, and 15 VOOs; Calabria (CAL) 13, 3, and 9 VOOs; and Sicily (SIC) 31, 30, and 18 VOOs.

In general, δ^2H followed a similar trend to that for $\delta^{13}C$: the δ^2H values increased for VOOs from the north to the south of Italy. δ^2H are affected by the isotopic composition of the primary hydrogen source (that is, water incorporated through the leaves), geographical, and climatic factors, as well as the biosynthetic pathways (*49*). As observed for $\delta^{13}C$, VOOs from northern regions (LOMB, TAA, VEN) presented the highest values of δ^2H for the three harvests studied, except for VOOs from TAA in the harvest 2006/07. This was probably due to the low rainfall and high mean temperatures in this region in 2006 (*50*). It should be noted that low δ^2H values for all three harvests were found for the VOOs from Abruzzo, a region in central Italy; δ^2H were similar to those for VOOs from northern regions of Italy. The highest δ^2H values for the three harvests were shown for VOOs from southern regions, that is, Calabria and Sicily. Campania for harvest 2004/05 and Liguria for all three harvests produced VOOs with relatively high δ^2H values. Liguria is in the Northwest of Italy (Figure 3), so lower δ^2H values were expected for its VOOs. However, several different factors influenced the $^2H/^1H$ ratios of Ligurian VOOs (*38, 40*); not only the latitude of the olive groves, but also the distance of plantations from the sea (Liguria is a narrow region between the mountains and the sea), and particular environmental conditions (microclima). All these factors made Ligurian VOOs different from those of the other northern regions (LOMB, VEN, TAA).

By grouping the Italian regions into four clusters on the basis of their latitude and the similarity of their isotopic ratios: North (Lombardia, Trentino Alto Adige, Veneto), Centre (Liguria, Umbria, Abruzzo, Lazio, Molise), South-1 (Campania, Puglia), and South-2 (Calabria, Sicily), as suggested by Camin et al. (*40*), $\delta^{13}C$ values of VOOs followed a similar trend (Figure 5) to that previously observed by those authors. (*40*). The median isotopic values for $\delta^{13}C$ of VOOs from the Centre and South-1 groups are similar for all the harvests studied. In contrast, the median values observed for the samples coming from the South-2 area were the highest values, and those from northern regions of Italy were the lowest ones, being quite different from those of Centre and South-1 macroareas, except for harvest 2006/07. This result may be due to the low rainfall before harvesting in the north during 2006 (*50*). According to the Standardized Precipitation Index (SPI), that measures the relative scarcity of water in one place, the rainfall in 2006 in the northeast of Italy (affecting the samples from the harvest 2006/07) was two points lower than the historical average rainfall data (*50*). Furthermore, the general low levels of rainfall throughout the country in 2006 caused the range of $\delta^{13}C$ values for VOOs from the harvest 2006/07 to be narrower.

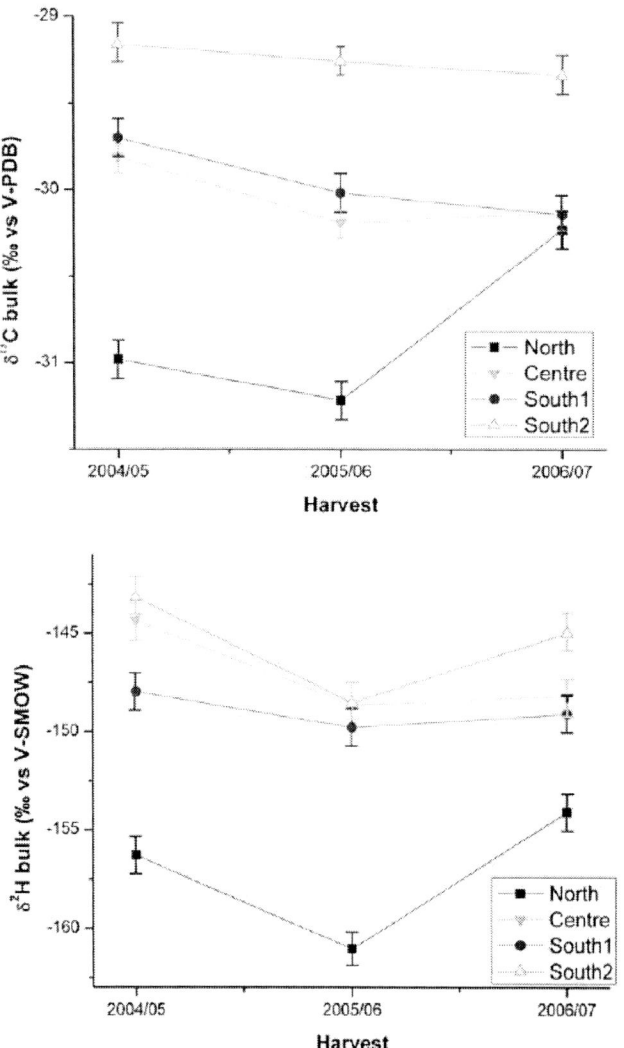

Figure 5. $\delta^{13}C$ and δ^2H median values for Italian VOOs grouped in macro-areas in three harvests (2004/05, 2005/06, and 2006/07): North (Lombardia, Trentino Alto Adige, Veneto), Centre (Liguria, Umbria, Abruzzo, Lazio, Molise), South-1 (Campania, Puglia), and South-2 (Calabria, Sicily).

As observed for $\delta^{13}C$, δ^2H values for VOOs from the North group were the lowest (Figure 5). In this case, the δ^2H values of VOOs from the Centre, South-1, and South-2 areas were closer to each other than the $\delta^{13}C$ values. This data showed the complex link between the various climatic and environmental factors which affect these δ-values. δ^2H values of VOOs are lower (more negative) for the years with higher rainfall and lower temperatures, as year 2005 was, affecting harvest 2005/06; and higher (more positive) for the years with higher temperatures and lower rainfall, as year 2006 was, influencing harvest 2006/07 (*50*). The range of δ^2H values of VOOs for the harvest 2006/07 was narrower than the other two harvests; not only did the higher temperatures in 2006 affect this but also the exceptionally dry conditions throughout the northwest of Italy that year. This made the isotopic values of the North group to be closer to the others.

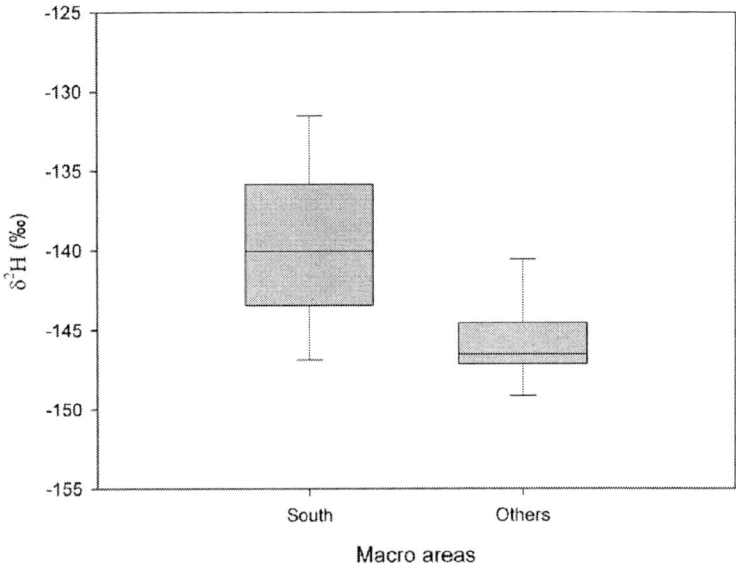

Figure 6. δ^2H median values of Spanish VOOs grouped in macro-areas in the harvest 2004/05: South (Andalusia and Extremadura, n = 42) and Others (other Spanish regions, n = 17).

3.3.3. Spanish Virgin Olive Oils

Ninety percent of Spanish VOO production is located in the south of Spain, that is, Andalusia. The climate in Andalusia depends very much on the geographical position: some inland areas are dry and cold while others are dry and hot, whereas the coastal areas are hot and humid. This is reflected in the large variability of the isotopic data of Andalusian VOOs. VOOs from Andalusia and Extremadura (southern regions) were compared to other Spanish VOOs from other regions for the harvest 2004/05 (only in this harvest enough samples were available to make this comparison). No significant differences were found between the $\delta^{13}C$ of VOOs of the two groups. In contrast, δ^2H values of Spanish VOOs (Figure 6) followed a similar trend to that observed for Italian oils (Figure 5); northern regions present more negative values than southern ones.

3.3.4. PDO Virgin Olive Oils

In Liguria (Italy) there is only one PDO covering the entire territory of production of VOOs called: *Riviera Ligure*. This PDO is divided in three geographical sub-denominations, namely *"Riviera dei Fiori"*, *"Riviera del Ponente Savonese"*, and *"Riviera di Levante"*, which correspond to the political provinces of Imperia, Savona, and Genoa/La Spezia, respectively (**Figure 7**). The province of Imperia has the largest production, while in Savona, Genoa, and La Spezia the production is located in a strip parallel to the sea. The factors affecting the isotopic ratios of VOOs from the PDO *Riviera* Ligure are: latitude, distance from the sea, cultivars, temperature, and rainfall. No significant differences in the δ^2H values of VOOs from the three geographical sub-denominations for the three harvests were observed (Table 3), as might be expected considering their similar climatic and geographical characteristics. $\delta^{13}C$ and δ^2H values of VOOs of the harvest 2004/05 were slightly higher than those from the two other harvests. In 2004, according to the meteorological data (*50*), a lower level of rainfall was recorded in Liguria compared to 2005 and 2006. $\delta^{13}C$ values of VOOs coming from *"Riviera dei Fiori"* (Imperia) showed the highest values for the three harvests. These results may be explained by the fact that Imperia is situated in southern latitude than the other Ligurian provinces.

In Sicily there are 6 PDOs for VOOs: two of them (*Monte Etna* and *Monti Iblei*) are located in the mountains on the eastern part of the island, and the other four (*Val di Mazara, Valdemone, Valle di Belice,* and *Valli Trapanesi*) comprise the valleys in the central and western part of the island. $\delta^{13}C$ values of the VOOs (**Table 4**) did not differ significantly between oils from the mountain areas and those from the valleys. However, δ^2H values of VOOs from the valleys were slightly lower than those from the mountains for the oils from the harvests 2004/05 and 2005/06. No differences were noted for the δ^2H values for the oils from the harvest 2006/07 between VOOs produced in the mountain and valley areas. The lower level of rainfall in Sicily in 2006 (*50*) was responsible for that fact.

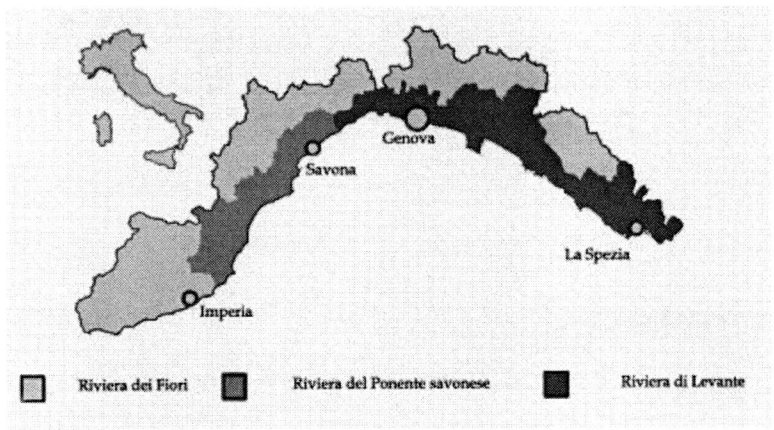

Figure 7. PDO *Riviera Ligure* and its geographical sub-denominations.

3.4. GEOGRAPHICAL CHARACTERIZATION OF VIRGIN OLIVE OIL

The large dataset of VOOs was studied regarding the situations that the antifraud authorities and regulatory bodies face. The PDO *Riviera Ligure*, some Italian regions, and the main VOO producing countries were used as examples to prove the potential of the tools to detect the mislabeling of non-PDO oils as PDO VOOs or the mislabeling of the provenance of VOOs at the regional or national level. With this purpose in mind, several multivariate data analysis techniques, datasets, types of data scaling, and crossvalidation were used. The best classification models were determined for each case study.

Table 3. Median, minimum, and maximum values of $\delta^{13}C$ and δ^2H of Ligurian VOOs from three harvests (2004/05, 2005/06, and 2006/07)

Harvest			2004/05				2005/06				2006/07		
Zone	Parameter	N	Median	Min.	Max.	N	Median	Min.	Max.	N	Median	Min.	Max.
Riviera di Levante	$\delta^{13}C$	10	-30.1	-30.6	-29.3	22	-30.7	-31.1	-29.4	25	-30.1	-31.5	-29.5
	δ^2H		-142	-146	-138		-149	-157	-143		-148	-157	-139
Riv. Ponente Savonese	$\delta^{13}C$	25	-29.9	-30.6	-28.9	22	-30.1	-30.9	-29.3	16	-30.4	-31.0	-29.5
	δ^2H		-144	-151	-132		-148	-156	-143		-147	-156	-137
Riviera dei Fiori	$\delta^{13}C$	28	-29.5	-30.1	-27.6	35	-30.0	-30.9	-28.8	27	-30.1	-31.2	-29.1
	δ^2H		-141	-148	-129		-146	-153	-139		-146	-153	-138

Table 4. Median, minimum, and maximum values of $\delta^{13}C$ and δ^2H of Sicilian VOOs from three harvests (2004/05, 2005/06, and 2006/07)

Harvest			2004/05				2005/06				2006/07		
Zone	Parameter	N	Median	Min.	Max.	N	Median	Min.	Max.	N	Median	Min.	Max.
Mountains	$\delta^{13}C$	17	-28.6	-29.8	-28.0	15	-29.4	-29.9	-28.5	9	-29.3	-30.6	-28.5
	δ^2H		-140	-151	-132		-145	-154	-139		-146	-158	-139
Valleys	$\delta^{13}C$	11	-29.2	-29.5	-28.8	15	-29.0	-29.9	-28.7	9	-29.3	-30.1	-28.6
	δ^2H		-144	-148	-142		-150	-155	-139		-145	-151	-143

3.4.1. PDO Olive Oil of *Riviera Ligure*

Under the PDO of *Riviera Ligure*, extra virgin olive oils produced in Liguria (Italy) that fulfill the PDO requirements related to olive varieties, farming practices, oil extraction procedures, bottling, and labeling (Dossier Number: IT/PDO/0017/1540, *Official Journal of the European Communities* 1997, L22) can be marketed. The ^{1}H NMR dataset of VOOs from different geographical origins and PDOs was used to differentiate between VOOs which did and did not belong to the PDO *Riviera Ligure*. Each VOO was represented in the 342-dimensional space by a data vector made of the 342 NMR variables. Univariate techniques (ANOVA, Fisher index, and box–whisker plots) cannot select a single variable to distinguish between Ligurian (belonging to the PDO *Riviera Ligure*) and non-Ligurian (not belonging to the PDO) samples. Therefore, it was necessary to apply supervised pattern recognition methods to build classification models that can distinguish VOOs of this PDO from the rest. Several multivariate approaches (LDA and PLS-DA) were tested using balanced or unbalanced data sets, different crossvalidation methods (LOO and 3-fold CV), and different data scalings (autoscaling and Pareto-scaling) to find the best approach for the authenticity and traceability of PDO olive oils. **Tables 5** and **6** summarize the classification results.

LDA models obtained using an unbalanced training-test set were biased to the class with more representatives, that is the non-Liguria class, presenting 93% of hits, whereas for the Liguria oils, the recognition and prediction abilities were <57% (**Table 5**). LDA was very sensitive to imbalances in the number of samples of each category in the dataset, as expected from the literature (*43*). In contrast, PLS-DA was not so sensitive to imbalances in the data set and performed better than LDA. The PLS-DA model using the autoscaled data set (three PLS components and the boundary at 0.318) presented recognition and prediction abilities in the crossvalidation of 88 and 80%, repetively for the VOOs from Liguria; and 83 and 84%, repetively for the non-Liguria VOOs. The percentage of classification of the final model (3 PLS components and the boundary at 0.318) and the prediction in the external validation were close to each other, 87 and 85% for Liguria VOOs, and 84 and 83% for non-Liguria VOOs, respectively; as well as to the recognition and prediction abilities in the training step. This would be considered indicative of a satisfactory model. However, the fact that the prediction ability was slightly higher than the recognition ability for the non-Liguria class indicated that the model did not perform properly for this class, which was probably due to the imbalances in

Table 5. Classification results obtained by supervised pattern recognition techniques for the authentication of VOO of the PDO *Riviera Ligure* using ^{1}H NMR spectral data (unbalanced data set) and δ^{13}C and δ^2H data[a]

Technique	Miscellaneous	Crossvalidation					Model					External Validation	
		Validation	N	prior prob	% Recognition		% Prediction		N	% Classification			% Prediction
					Liguria	Non-Liguria	Liguria	Non-Liguria		Liguria	Non-Liguria	Liguria	Non-Liguria
			Liguria:126; Non-Liguria: 466	Liguria: 0.21; Non-Liguria: 0.79					126	0.21	0.79	73	270
LDA[b]	5 NMR buckets selected: 6.61, 5.09, 4.57, 4.05 and 0.33 ppm; autoscaling	3-fold CV			56.7	93.5	54.0	93.3		56.3	93.6	45.2	92.6
LDA[b]	5 NMR buckets selected: 6.61, 5.09, 4.57, 4.05 and 0.33 ppm; autoscaling IRMS: δ^{13}C and δ^2H	3-fold CV			56.7	93.5	54.0	93.3		57.9	94.4	45.2	92.6
PLS-DA[b]	3 PLS components selected boundary: 0.3180; autoscaling	3-fold CV			87.7	83.4	80.2	84.3		86.5	83.9	84.9	83.3
PLS-DA[c]	3 PLS components selected boundary: 0.3180; autoscaling	3-fold CV			-	-	81.7	84.1		86.5	83.7	84.9	83.3

Table 5. (Continued).

Technique	Miscellaneous	Crossvalidation		Model				External Validation	
		Validation	% Recognition		% Prediction		% Classification		% Prediction
			Liguria	Non-Liguria	Liguria	Non-Liguria	Liguria	Non-Liguria	Liguria	Non-Liguria
	N		Liguria:126; Non-Liguria: 466				126	466	73	270
	prior prob		Liguria: 0.21; Non-Liguria: 0.79				0.21	0.79		
PLS-DA[c]	3 PLS components selected boundary: 0.3180; autoscaling	LOO	-	-	81.7	83.0	86.5	83.7	84.9	83.3
PLS-DA[d]	3 PLS components selected boundary: 0.3180; autoscaling	3-fold CV/ LOO	-	-	-	-	86.5	83.9	84.9	83.3
PLS-DA[d]	3 PLS components selected boundary: 0.3175; Pareto scaling	3-fold CV/ LOO	-	-	-	-	84.1	81.1	74.0	77.8

[a] Abbreviations: N, number of samples; prior prob, prior probability; LDA, linear discriminant analysis; PLS-DA, partial least square discriminant analysis; Class codes: Liguria, 1; non-Liguria, 0.
[b] Statistica.; [c] The Unscrambler.; [d] SIMCA-P.

Table 6. Classification results obtained by supervised pattern recognition techniques for the authentication of VOO of the PDO *Riviera Ligure* using ^{1}H NMR spectral data (balanced data set) and δ^{13}C and δ^2H data[a]

Technique	Miscellaneous	Crossvalidation					Model				External Validation	
		Validation	N		% Recognition		% Prediction		N	% Classification		% Prediction
				prior prob	Liguria	Non-Liguria	Liguria	Non-Liguria		Liguria	Non-Liguria	Liguria	Non-Liguria
			Liguria:132; Non-Liguria: 135	Liguria: 0.49; Non-Liguria: 0.51					132	135	0.49	0.51	67	601
LDA[b]	5 NMR buckets selected: 6.61, 5.11, 4.57, 4.05 and 0.33 ppm; autoscaling	3-fold CV			84.1	85.9	84.1	83.7	82.6	85.2	86.6	79.7
LDA[b]	4 NMR buckets selected: 5.11, 4.57, 4.05 and 0.33 ppm	3-fold CV			88.3	84.1	85.6	80.7	87.9	83.0	89.6	79.7
PLS-DA[b]	IRMS: δ^{13}C and δ^2H	3-fold CV			91.3	92.6	87.9	86.7	91.7	90.4	88.1	85.5
PLS-DA[c]	5 PLS components selected boundary: 0.540; autoscaling	3-fold CV			-	-	86.4	85.9	91.7	90.4	88.1	85.5
PLS-DA[c]	5 PLS components selected boundary: 0.540; autoscaling	3-fold CV			-	-	86.4	85.2	91.7	91.9	86.6	86.0
PLS-DA[c]	5 PLS components selected IRMS: δ^{13}C and δ^2H boundary: 0.547; autoscaling	3-fold CV			-	-	87.1	85.9	91.7	90.4	88.1	85.5
PLS-DA[c]	5 PLS components selected boundary: 0.540; autoscaling	LOO			-	-						

Table 6. (Continued).

Technique	Miscellaneous	Crossvalidation					Model		External Validation	
		Validation	N	prior prob	% Recognition		% Prediction		% Classification		% Prediction	
					Liguria	Non-Liguria	Liguria	Non-Liguria	Liguria	Non-Liguria	Liguria	Non-Liguria
PLS-DA [c]	5 PLS components selected IRMS: $\delta^{13}C$ and δ^2H boundary: 0.547; autoscaling	LOO	Liguria:132; Non-Liguria: 135	Liguria: 0.49; Non-Liguria: 0.51	-	-	85.6	85.2	132	135	67	601
									0.49	0.51		
PLS-DA [d]	5 PLS components selected boundary: 0.540; autoscaling	3-fold CV/ LOO	-	-	-	-	-	-	91.7	90.4	88.1	85.5
PLS-DA [d]	4 PLS components selected boundary: 0.520; Pareto scaling	3-fold CV/ LOO	-	-	-	-	-	-	91.7	91.9	86.6	86.0
									87.1	83.0	80.6	81.0

[a] See abbreviations: Table 5.
[b] Statistica.; [c] The Unscrambler.; [d] SIMCA-P.

the data set. Indeed, the percentage of correct classifications for the non-Liguria category was slightly higher than its *a priori* probability (79%). The crossvalidation used in the training step did not influence the prediction abilities for either category, indicating that the samples were well-represented in the training set. PLS-DA using the autoscaled unbalanced data set attained the same best final model, consisting of three PLS components and the boundary at 0.318 (class codes: Liguria, 1; non-Liguria, 0). PLS-DA applied on the Pareto-scaled unbalanced data set provided a model with 4 PLS components and the boundary at 0.325, which performed worse.

Both supervised pattern recognition techniques, LDA and PLS-DA, performed better if a balanced training-test set was used (**Table 6**). However PLS-DA still outperformed LDA. LDA achieved classifications of around 85% of hits for both categories. PLS-DA provided a model with five PLS components and the boundary at 0.540, which achieved slightly better results for the Liguria class (prediction ability in the crossvalidation, 86-88%; classification ability of the final model, 92%; and prediction ability of the final model in the external validation, 88%) than for the non-Liguria VOOs (86-87, 90, and 86%, respectively). These results together with the facts that in the crossvalidation the recognition ability was higher but close to the prediction ability and the classification ability of the final model was also higher but close to prediction ability in the external validation, disclosed that the model achieved was feasible and not random, as well as well-represented by the samples in the dataset. PLS-DA on the Pareto-scaling balanced data set produced a model with four PLS components and the boundary at 0.520. This model also gave better classifications (classification ability of the final model, 87 and 83% for the Liguria and non-Liguria, repectively; and prediction ability of the final model in the external validation, 81% for both classes) than the model made with the unbalanced data set. With both data sets (balanced and unbalanced), Pareto-scaling led to worse outcomes than autoscaling.

The variable selection used for LDA afforded five NMR buckets centered at the following chemical shifts: 6.61, 5.11 or 5.09, 4.57, 4.05, and 0.33 ppm. These buckets correspond to signals of the following VOO components: phenolic compounds and unsaturated alcohols, which present characteristic resonances in the spectral region 6–7.5 ppm (*53*) and 4.5–5 ppm, repectively; *sn*-1,2-diglycerides (5.09–5.11 ppm) and *sn*-1,3-diglycerides (4.05 ppm), due to their CH glycerol protons; and cycloartenol (0.33 ppm), to the methylene proton of its cyclopropanoic ring (*47*).

The weighted regression coefficients of the PLS models indicate the importance of the NMR variables on the model: the larger the regression coefficient, the higher the influence of the variable on the PLS model (*54*). The variables selected in LDA were among the variables that presented the highest weighted regression coefficients in the PLS-DA models: 6.85–6.83, 6.75, 6.67, 6.59, and 6.23 ppm belonged to signals of phenolic compounds; 5.15–5.07 ppm was due to the CH glycerol protons of *sn*-1,2-diglycerides; 4.99 ppm was due to unsaturated alcohols; 4.71, 4.65, and 4.57 ppm, were due to terpenes; 2.79 ppm was due to diallylic proton of linolenic acyl group; 1.29 ppm was due to methylene proton of linoleic and linolenic acyl group; and 0.33 ppm was due to cycloartenol. Therefore, both pattern recognition techniques arrived at consistent results, each one providing information about the most important features for the characterization of PDO *Riviera Ligure* VOOs.

With the additional information provided by the $\delta^{13}C$ and $\delta^{2}H$ isotopes measured by IRMS, the classification results of the LDA model using the unbalanced dataset were similar, even though both isotopes were significant variables (at the 5% level) together with the five NMR buckets previously selected (**Table 5**). This was probably due to the imbalances in the dataset. However, with the balanced data set, LDA provided a model with four significant NMR variables previously selected and the two isotopes (both significant) that obtained (**Table 6**): (*i*) slightly better classifications for the Liguria class than the model made only with NMR data; (*ii*) whereas the classification results for the non-Liguria class were slightly worse. This fact suggested that $\delta^{13}C$ and $\delta^{2}H$ isotopes contained some information related to VOOs from Liguria. As a matter of fact, Angerosa et al. (*38*) and, more recently, Camin et al. (*40*) observed that the values of $\delta^{13}C$ and $\delta^{2}H$ in olive oils increased according to the olive cultivation latitude, from northern to the southern Italy. Thus, olive oils produced in the northern, the central and the southern Italy could be well differentiated by these isotopes. However, the discrimination of olive oils produced in different regions at similar latitudes is more difficult. Olive oils from Liguria (in northwestern Italy) present a different chemical composition and particular organoleptic characteristics, in comparison with other northern Italian olive oils such as those from the region of lake Garda (Lombardia and Veneto). This is due to the proximity of Liguria to the sea and the special climate of the region. PLS-DA was also applied on an autoscaled balanced data set that contained 344 variables (342 NMR buckets and the 2 isotopes); however, no improvement was observed in the classification results of the models achieved (neither LOO nor 3-fold CV). In conclusion, the best model for the distinction between VOOs belonging to the PDO *Riviera Ligure* and other VOOs was afforded by PLS-DA using an autoscaled balanced training-test set.

3.4.2. Virgin Olive Oils from Italian Regions

The large data set available (963 × 342 matrix) was used to authenticate VOOs produced in selected Italian regions. The regions selected were those best represented in the dataset. Models were generated for the following Italian regions: Umbria (which is also a registered PDO: PDO *Umbria*, Dossier Number: IT/PDO/0017/1520, *Official Journal of the European Communities* 1997, L322 25.11.1997); Sicily (six PDOs: *Monte Etna, Val di Mazara, Valli Trapanesi, Valle del Belice, Valdemone,* and *Monti Iblei*), Puglia (four PDOs: *Terra d'Otranto, Collina di Brindisi, Dauno,* and *Terra di Bari*), Lazio (three PDOs: *Tuscia, Canino,* and *Sabina*), Garda (three PDOs: *Garda, Laghi Lombardi,* and *Veneto Valpolicella, Veneto Euganei e Berici, Veneto del Grappa*), Campania (three PDOs: *Peninsola Sorrentina, Colline Salernitane, and Cilento*), and Calabria (three PDOs: *Lametia, Alto Crotonese, and Bruzio*).

Taking into account the results obtained by the different approaches studied for the PDO *Riviera Ligure*, and that the number of samples for each of these regions was considerably smaller than in the Liguria category, the models for these regions were developed using an autoscaled balanced training-test set by PLS-DA and LOO CV. The final models were also evaluated by external validation. The results are summarized in **Table 7**. The model obtained to authenticate VOOs from Sicily recognized 98% of the Sicilian oils and 89% of the non-Sicilian ones, and managed to correctly predict in the crossvalidation step 93 and 86% of Sicilian and non-Sicilian oils, respectively. Since this model achieved similar predictions in the external validation (>85% of hits for both categories) to those in the modeling step, it can be considered stable and robust. In contrast, the models created for other regions such as Lazio, Garda, and Calabria were not so satisfactory: although the classification abilities were close to 90% of correct hits or even higher, the prediction abilities in the crossvalidation were from 10% to 24% lower, which meant that the classification results were very dependent on the samples included in the training set in the modeling step. This also occurred for Umbria and Campania, but the models achieved about 80% of correct classification for the training set, and predictions on the test set were more than 10% lower, except for the oils belonging to the non-Umbria category (5% less). The external validation of some models (only 50% of Umbria, 57% of Campania, and 60% of Calabria VOOs were correctly predicted) confirmed that the classes were not well represented in the modeling step. Puglia VOOs, as well as non-Garda VOOs, were much

Table 7. Classification results obtained by supervised pattern recognition techniques for the authentication of VOO from certain Italian regions using ^{1}H NMR spectral data and δ^{13}C and δ^2H data[a]

		Crossvalidation					Model					External Validation			
		% Prediction					% Classification					% Prediction			
Model	Origin	N	prior prob	a) NMR	b) NMR + IRMS	N	prior prob	NMR	NMR + IRMS	N	prior prob	a) NMR	b) NMR + IRMS
Umbria vs. non-Umbria	Umbria	35	0.45	71.4	71.4	35	0.45	82.9	82.9	12	0.014	50.0	50.0
a) NMR: 2 PLS components, boundary: 0.525	Non-Umbria	43	0.55	74.4	72.1	43	0.55	79.1	79.1	845	0.986	74.8	76.7
b) NMR+IRMS: 2 PLS components, boundary: 0.5325													
Sicily vs. non-Sicily	Sicily	54	0.47	92.6	92.6	54	0.47	98.1	98.1	24	0.029	87.5	91.7
a) 3 PLS components, boundary: 0.460	Non-Sicily	62	0.53	85.5	85.5	62	0.53	88.7	87.1	795	0.971	85.8	84.9
b) NMR+IRMS: 3 PLS components, boundary: 0.452													

Model			Crossvalidation				Model			External Validation			
				% Prediction					% Classification			% Prediction	
Model	Origin	N	prior prob	a) NMR	b) NMR + IRMS		N	prior prob	NMR	NMR + IRMS	N	prior prob	a) NMR	b) NMR + IRMS
Puglia vs. non-Puglia														
a) 2 PLS components, boundary: 0.4435	Puglia	47	0.42	68.1	63.8		47	0.42	72.3	72.3	22	0.027	81.8	81.8
b) NMR+IRMS: 2 PLS components, boundary: 0.451	Non-Puglia	64	0.58	62.5	70.3		64	0.58	71.9	71.9	802	0.973	65.1	67.1
Lazio vs. non-Lazio														
a) 4 PLS components, boundary: 0.515	Lazio	40	0.49	80.0	85.0		40	0.49	97.5	97.5	19	0.022	73.7	84.2
b) NMR+IRMS: 4 PLS components, boundary: 0.533	Non-Lazio	41	0.51	68.3	75.6		41	0.51	90.2	90.2	835	0.978	69.3	70.5
Garda vs. non-Garda	Garda	36	0.46	72.2	77.8		36	0.46	91.7	88.9	13	0.015	69.2	76.9
a) 3 PLS components, boundary: 0.555	Non-Garda	43	0.54	74.4	74.4		43	0.54	90.7	90.7	843	0.985	80.1	81.9

Table 7. (Continued).

Model		Crossvalidation				Model				External Validation			
		% Prediction				% Classification				% Prediction			
	Origin	N	prior prob	a) NMR	b) NMR + IRMS	N	prior prob	NMR	NMR + IRMS	N	prior prob	a) NMR	b) NMR + IRMS
Campania vs. non-Campania													
a) 2 PLS components, boundary: 0.430 b) NMR+IRMS: 3 PLS components, boundary: 0.538	Campania	21	0.43	71.4	71.4	21	0.43	81.0	81.0	7	0.008	57.1	57.1
	Non-Campania	28	0.57	64.3	64.3	28	0.57	78.6	78.6	879	0.992	62.9	63.0
Calabria vs. non-Calabria													
a) 3 PLS components, boundary: 0.4315 b) NMR+IRMS: 2 PLS components, boundary: 0.445	Calabria	17	0.38	70.6	70.6	17	0.38	94.1	94.1	5	0.006	60.0	60.0
b) NMR+IRMS: 3 PLS components, boundary: 0.447	Non-Calabria	28	0.62	85.7	85.7	28	0.62	96.4	96.4	885	0.994	79.9	80.3

[a] See abbreviations: Table 5; Models obtained by PLS-DA using autoscaling, LOO and The Unscrambler; Class codes: "Region", 1; "non-Region, 0.

better predicted in the external data set (82% of hits) than in the crossvalidation (68 and 72% of hits, respectively). This was probably due to the way samples were divided into the training-test set and the external set: the PCA scores of all the VOOs were regarded to select samples from the whole cloud of points including the borders. This procedure ensured that the training-test set was representative of all the samples (at least of the three harvests studied); however, the predictions on the external set could be overoptimistic.

The most influential variables, that is, those with the highest weighted regression coefficients, on the binary PLS-DA models achieved for each region are listed in Table 8. The signals due to cycloartenol (0.31–0.33 ppm) and *sn*-1,2-diglycerides (5.07–5.15-ppm) were important for all models except for Garda, as well the resonances in the phenolic region at 6.73–6.79 ppm, which only did not influence the model for Sicily. The acyl group methylene protons of saturated fatty acids (1.23 ppm), ^{13}C satellite of signal at 4.09–4.32 ppm (α-methylene protons of the glyceryl group of triglycerides) at 3.97 ppm and the signal at 5.57 ppm were important specifically for the Umbria model; the signals at 0.53 ppm and 0.79 ppm, for the Sicily model; the methylic proton of the C18-steroid group of β-sitosterol (0.67 ppm) and the terpene signal at 4.57–4.59 ppm, for the Puglia model; the signal of the cycloartenol at 0.55 ppm, ^{13}C satellite of signal at 2.26–2.32 ppm (α-methylene protons of the acyl group) at 2.15 ppm, the glycerol proton of *sn*-1,2-diglycerides (3.71 ppm) and signals at 6.19 ppm and 6.15 ppm in the phenolic region, for the Lazio model; signals in the regions 1.35–1.43 ppm, 2.35–2.39 ppm, and 4.33–4.35 ppm, the α-methylene protons of the acyl group (2.29 ppm and 2.33 ppm), the signal at 3.75 ppm, the α-methylene protons of the glyceryl group of triglycerides (4.27 ppm), and the signal at 6.15 ppm in the phenolic region for the Campania model; and the signal at 5.93 ppm for the Calabria model. The glyceryl protons of *sn*-1,3-diglycerides (4.05–4.07 ppm), and triglycerides (5.25 and 5.29 ppm) were influential for the models of Umbria, Lazio, Umbria, and Campania, repectively; signals in the phenolic region at 6.25–6.29 ppm for the models of Puglia and Calabria; signals in the phenolic region at 6.63–6.65 and 6.69–6.71 ppm for the models of Umbria and Campania; and signals in the phenolic region at 6.45–6.47 ppm for the models of Umbria and Garda.

Table 8. Most important variables in the binary PLS-DA models achieved for the geographical classification of VOOs at the regional level by ^{1}H NMR

Umbria	Sicilia	Puglia	Lazio	Garda	Campania	Calabria	Functional group	Attribution
0.33	0.33-0.31	0.31	0.33-0.31	0.29		0.33-0.31	-CH_2- (cyclopropanic ring)	Cycloartenol
	0.53		0.55				-CH_2- (cyclopropanic ring)	Cycloartenol
		0.67					-CH_3 (C18-steroid group)	β-sitosterol
	0.79							
	0.95		0.97		0.95-0.91	0.97	-CH_3 (acyl group)	linolenic (or ω-3)
1.01	1.03-0.99	1.01-0.99					-CH_3 (^{13}C satellite of signal at 0.87 ppm)	saturated (palmitic, stearic)
1.23							-(CH_2)$_n$- (acyl group)	
1.27		1.25	1.25	1.27	1.25		-(CH_2)$_n$- (acyl group)	Oleic
1.29			1.29		1.29	1.29	-(CH_2)$_n$- (acyl group)	linoleic and linolenic
					1.43-1.35			
1.67, 1.61-1.59	1.67, 1.59	1.67			1.65, 1.59, 1.51		-OCO-CH_2-CH_2- (acyl group)	
			1.75-1.73		1.75-1.73			
	2.07-2.03, 1.99		1.99, 1.95	2.05, 2.01	2.05-1.99		-CH_2-CH=CH- (acyl group)	
			2.15				-OCO-CH_2- (^{13}C satellite of signal at 2.26-	

Umbria	Sicilia	Puglia	Lazio	Garda	Campania	Calabria	Functional group	Attribution
					2.33, 2.29		2.32 ppm, acyl group)	
					2.39-2.35		-OCO-CH_2- (acyl group)	
2.75-2.73	2.75-2.71			2.75-2.73			=CH-CH_2-CH= (acyl group)	Linoleic
2.77	2.77			2.77	2.77		=CH-CH_2-CH= (acyl group)	linoleic and linolenic
	2.79			2.79	2.83	2.81-2.79	=CH-CH_2-CH= (acyl group)	Linolenic
							-CH_2OH (glyceryl group)	sn 1,2-diglycerides
3.97			3.71		3.75		-CH_2OCOR (^{13}C satellite of signal at 4.09-4.32 ppm, glyceryl group)	Triglycerides
4.07-4.05			4.07-4.05				>CH-OH (glyceryl group)	sn 1,3-diglycerides
					4.27		-CH_2OCOR (glyceryl group)	Triglycerides
		4.59-4.57			4.35-4.33			Terpene
4.65	4.65	4.65	4.65			4.65		Terpene
4.71-4.69	4.71-4.69	4.71	4.71					Terpene
5.15-5.07	5.15-5.07	5.15-5.07	5.15-5.11		5.15-5.07	5.15-5.07	>CHOCOR (glyceryl group)	sn 1,2-diglycerides

Table 8. (Continued).

Umbria	Sicilia	Puglia	Lazio	Garda	Campania	Calabria	Functional group	Attribution
5.29					5.25		>CHOCOR (glyceryl group)	Triglycerides
5.33	5.37		5.33		5.43-5.33	5.33	-CH=CH- (acyl group)	
5.57								
5.75	5.73					5.93		
6.01-5.97	5.99			6.01-5.97				phenolic compounds
6.05-6.03			6.05-6.03	6.05				
			6.19, 6.15					
		6.29-6.25	6.23	6.23		6.23-6.21		
					6.37	6.29-6.25		
6.47-6.45				6.47				
6.55-6.53	6.55-6.53			6.57-6.51				
6.61		6.59	6.59	6.61-6.59	6.61-6.59			
6.65-6.63					6.63			
6.71-6.69					6.71-6.67			
6.75		6.77-6.73	6.77-6.75	6.79-6.73	6.79-6.73	6.77		
6.95			6.95-6.85	6.97-6.85		6.87		

These results disclosed that ^{1}H NMR spectra of VOOs contained information related to the region of provenance of the oil, but further studies should be carried out with a considerably larger sample set for each region, and even for each of their PDOs to guarantee the detection of fraud when VOO is falsely labeled as belonging to a certain origin. In this regard, Sicily, which is an island at the southernmost point of Italy, produces an olive oil that is markedly influenced by pedoclimatic factors, in accordance with its geographical position. It is therefore coherent that the VOO produced on this island presents a characteristic chemical composition that allows one to distinguish it from all other VOOs from different geographical regions. In contrast, the stable isotopes measured on the samples, δ^{13}C and δ^2H, which are commonly related to pedoclimatic features, did not improve the classification of the Sicilian and non-Sicilian VOOs, even if δ^{13}C was a significant variable in the model. Similarly, these stable isotopes did not enhance the classification results for the Umbria, Campania, and Calabria models. In the Umbria model, both isotopes were among the variables with the highest weighted regression coefficients, whereas they were not in the Campania and Calabria models. However, both stable isotopes provided extra useful information related to the non-Puglia category in the Puglia model, as well as for the Garda class in the Garda model. δ^{13}C data substantially improved the classification for both classes of the Lazio model. Therefore, depending on the case study, stable isotopes can provide useful and complementary information to that contained in the ^{1}H NMR fingerprint of the VOOs related to their geographical origin. Camin et al. (*40*) found that olive oils produced in the southern regions of Italy had similar isotopic signatures, making a clear discrimination among them difficult. The improvement observed on the Garda model can be attributed to the fact that these isotopes contain information referring to the latitude of this region, that is, the northern part of Italy, and its special environmental conditions, that is, low mean temperatures and rainy weather. Thus, Garda olive oils present lower δ^{13}C and δ^2H than olive oils produced in the rest of Italy.

3.4.3. The Main Producers of Virgin Olive Oil: Spain, Italy, and Greece

With regard to the adulteration of VOOs from a certain country with VOOs produced in another country at a lower cost, or the false labeling of the VOOs as coming from a certain country when they were actually produced in another, the need for chemical approaches to detect these fraudulent activities is evermore apparent. The ^{1}H NMR data of the VOOs from the main olive oil producing countries, that is, Spain, Italy, and Greece, were analyzed by multivariate techniques with the purpose of creating classification models that would allow the distinction between the geographical origins of VOOs from these three countries. **Table 9** shows the results. The model distinguishes VOOs from Greece from all the rest of the VOOs; it classified properly >97% of the samples of both categories, Greece and non-Greece, and predicted correctly >90% of the samples in the test set of the crossvalidation, as well as in the external validation. The binary models for Italy and Spain presented classification abilities of 89% for the Italian oils and the Spanish oils, 84% for the non-Italy category, and 85% for the non-Spain category. The prediction abilities in the crossvalidation for the model for Spain were ca. 80% of hits for both classes, whereas the predictions in the external validation were considerably different, for the Spanish VOOs it was overoptimistic (92%), and for the non-Spanish VOOs it was considerably low (67%). In the model for Spain, the variability of the non-Spain category was under-represented in the training-test sets. Therefore, this model did not provide good predictions for this category in the external set. The model for Italy provided prediction abilities in the crossvalidation of ca. 76% for both classes and in external validation, close to this value. These predictions were substantially lower than the classification ability of the model, indicating that the model was dependent on the samples included in the training set.

Table 10 gathers the most influential variables, that is, those with the highest weighted regression coefficients, on the binary PLS-DA models obtained for each country, identifying the functional groups and compounds to which the signals are due. The signals in the phenolic regions at 6.45–6.47 and 6.83–6.85 ppm were important for the three models. In contrast, the model for Spain was particularly influenced by the methylic proton of the C18-steroid group of β-sitosterol (0.67 ppm), the β-methylene protons of the acyl group (1.59 and 1.67 ppm), the allylic protons of the acyl group (1.99–2.07 ppm), the diallylic protons of the acyl group of linoleic (2.73–2.77 ppm), and linolenic (2.77–2.81 ppm), the glycerol proton of *sn*-1,2-diglycerides (3.71 ppm), *sn*-1,3-diglycerides (4.05–4.07 ppm), and triglycerides (5.25 ppm and 5.29 ppm), the olefinic protons of the acyl groups (5.37 ppm), the signals in the phenolic region at 6.37, 6.61 and 6.71 ppm, and the signals at 0.53, 1.75–1.77, and 2.35 ppm. Among the most important variables, those that affected only the model for Greece were the methylic proton of the linolenic acyl group (0.97 ppm), the terpene signal at

4.55–4.57 ppm, and the signals at 0.77 and 3.81 ppm. The resonances of cycloartenol (0.31–0.33 ppm and 0.55 ppm) and phenolic compounds at 6.23 and 6.27 ppm were important for the models of Italy and Greece.

A ternary model was developed to classify VOOs from three countries: Italy, Spain, and Greece. It did not classify as well as the binary models created for each country (**Table 9**). Indeed, PLS-DA is known to perform better with a smaller number of classes (*43*). Thus, the model recognized 94% of Greek, 81% of Italian, and 75% of Spanish oils in the training-test sets and predicted correctly 88% of Greek oils and 69% of the samples from Italy and Spain in the crossvalidation and 81% of Greek, 79% of Spanish, and 66% of Italian oils in the external validation.

In conclusion, these results show that ^{1}H NMR fingerprinting of VOOs can be a useful tool to ensure authenticity and traceability of VOOs at the national level. From this study, a stable model was achieved to distinguish Greek VOOs from oils from other countries. However, for Italian and Spanish VOOs further studies should be performed with a larger balanced data set, in which all categories will be well represented, to obtain robust models. In the present data set, Spain was clearly under-represented, being the main producer (50% of EU production of olive oil); for Italy, even though it was quite well-represented, the number of samples were very unbalanced with regard to the other countries and, therefore, few Italian samples were used in the modeling step, so the classification results were very dependent on the samples in the training-test set.

Stable isotopes considerably enhanced the classification results of VOOs according to their country of origin, $\delta^{13}C$ and δ^2H being among the highest weighted regression coefficients of the PLS-DA models. The results for the model for Italy provided better discrimination for both categories, whereas those of the model for Spain provided better classifications for the non-Spanish category. Higher prediction abilities for the Greek category were obtained in the crossvalidation; however, the percentage of classification of the final model and the predictions on the external data set did not improve the classification notably. In contrast, the additional information provided by these isotopes enhances substantially the classification results for the Greek and Italian categories of the ternary model afforded for the three countries. Indeed, $\delta^{13}C$ of olive oils already showed potential for the discrimination between olive oils from Greece, Spain, and Italy (*37, 38*), because the median value of $\delta^{13}C$ increases in the order Italian < Greek < (or ≅) Spanish oils.

Table 9. Classification results obtained by supervised pattern recognition techniques for the authentication of VOO from the main producing countries, that is, Italy, Spain, and Greece, using ^{1}H NMR spectral data and δ^{13}C and δ^2H data[a]

Model		Crossvalidation % Prediction				Model % Classification					External Validation[b] % Prediction			
	Origin	N	prior prob	a) NMR	b) NMR + IRMS	N	prior prob	NMR	NMR + IRMS	N	prior prob	a) NMR	b) NMR + IRMS
Italy vs non-Italy	Italy	72	0.35	75.0	79.2	72	0.35	88.9	90.3	568	0.78	75.7	80.1
a) NMR: 4 PLS components, boundary: 0.4020	Non-Italy	135	0.65	77.0	85.2	135	0.65	84.4	90.4	160	0.22	71.9	73.1
b) NMR+IRMS: 4 PLS components, boundary: 0.4225													
Spain vs non-Spain	Spain	71	0.34	78.9	77.5	71	0.34	88.7	87.3	70	0.10	92.9	90.0
a) NMR: 3 PLS components, boundary: 0.3563	Non-Spain	136	0.66	80.9	83.8	136	0.66	85.3	88.2	658	0.90	67.2	71.7
b) NMR+IRMS: 3 PLS components, boundary: 0.3677													
Greece vs non-Greece	Greece	64	0.31	92.2	96.9	64	0.31	98.4	98.4	31	0.04	96.8	96.8
a) NMR: 5 PLS components, boundary: 0.4725	Non-Greece	143	0.69	93.7	91.6	143	0.69	97.9	96.5	697	0.96	90.0	91.1

Model	Crossvalidation					Model %Classification				External Validation[b]			
	Origin	N	prior prob	% Prediction		N	prior prob	NMR	NMR+IRMS	N	prior prob	a) NMR	b) NMR+IRMS
				a) NMR	b) NMR+IRMS								
Italy vs Spain vs Greece													
a) NMR: 5 PLS components, boundary: 0.4120 b) NMR+IRMS: 5 PLS components, boundary: 0.4500	Italy	72	0.35	69.4	79.2	72	0.35	80.6	83.3	568	0.85	65.5	73.8
a) NMR: 5 PLS components, boundary: 0.3570 b) NMR+IRMS: 5 PLS components, boundary: 0.3616	Spain	71	0.34	69.0	67.6	71	0.34	74.6	77.5	70	0.10	78.6	78.6
b) NMR+IRMS: 5 PLS components, boundary: 0.4582													

Table 9. (Continued).

Model		Crossvalidation				Model				External Validation[b]			
				% Prediction				% Classification				% Prediction	
	Origin	N	prior prob	a) NMR	b) NMR + IRMS		N	prior prob	NMR	NMR + IRMS	N	prior prob	a) NMR	b) NMR + IRMS
a) NMR: 5 PLS components, boundary: 0.4395	Greece	64	0.31	87.5	90.6		64	0.31	93.8	100.0	31	0.05	80.6	83.9
b) NMR+IRMS: 5 PLS components, boundary: 0.4350														

[a] See abbreviations: **Table 5**; Models obtained by PLS-DA using autoscaling, LOO and The Unscrambler; Class codes: "Country", 1; "non-Country", 0.
[b] The external data set used to evaluate the 3-classes model consisted of samples from Italy, Spain, and Greece.

Table 10. Most important variables in the binary PLS-DA models achieved for the geographical classification of VOOs at the national level by ^{1}H NMR

Italy	Spain	Greece	Functional group	Attribution
0.33-0.31	0.33-0.31	0.33-0.31	-CH_2- (cyclopropanic ring)	cycloartenol
0.55	0.53	0.55	-CH_2- (cyclopropanic ring)	cycloartenol
	0.67		-CH_3 (C18-steroid group)	β-sitosterol
		0.77		
	1.67, 1.59	0.97	-CH_3 (acyl group)	linolenic (or ω-3)
	1.77-1.75		-OCO-CH_2-CH_2- (acyl group)	
	2.07-1.99		-CH_2-CH=CH- (acyl group)	
	2.35			
	2.75-2.73		=CH-CH_2-CH= (acyl group)	linoleic
	2.77		=CH-CH_2-CH= (acyl group)	linoleic and linolenic
	2.81-2.79		=CH-CH_2-CH= (acyl group)	linolenic
	3.61	3.63-3.61		
	3.71		-CH_2OH (glyceryl group)	sn 1,2-diglycerides
		3.81		
	4.07-4.05	4.57-4.55	>CH-OH (glyceryl group)	sn 1,3-diglycerides
	4.65	4.65		terpene
	4.71-4.69	4.71-4.69		terpene
	5.15-5.09	5.13	>CHOCOR (glyceryl group)	sn 1,2-diglycerides
	5.25		>CHOCOR (glyceryl group)	terpene
	5.37		-CH=CH- (acyl group)	triglycerides

Table 10. (Continued).

Italy	Spain	Greece	Functional group	Attribution
6.23	5.75-5.71	5.73		
6.27	6.45	6.23		
6.47-6.45	6.37	6.27		phenolic compounds
	6.45	6.47-6.45		
	6.61			
6.79	6.71			
6.85	6.79-6.73			
	6.85	6.85-6.83		

ACKNOWLEDGMENT

The authors thank the EU TRACE project for funding this work and the research groups that participated in the collection of the olive oil samples: Laboratorio Arbitral Agroalimentario (Ministry of Agriculture and Fishery, Spain), General Chemical State Laboratory D'xy Athinon (Greece), General State Laboratory (Ministry of Health, Cyprus), Departamento de Química Orgánica - Universidad de Córdoba (Spain), Istituto di Metodologie Chimiche (CNR, UNAPROL, Dipartimento di Chimica e Technologie Farmaceutiche ed Alimentari, Università di Genova, Italia), Fondazione Edmund Mach (Istituto San Michele all'Adige, Italy), and Eurofins Scientific Analytics (France). The authors would like to acknowledge K. Héberger and N. Segebarth for fruitful discussions.

The information contained in this book reflects the authors' views; the European Commission is not liable for any use of the information contained herein.

REFERENCES

[1] Harwood, J. L.; Aparicio, R., *Handbook of olive oil: analysis and properties*. Aspen: Gaithersburg, MD, 2000; p 620.
[2] International Oil Council. http://www.internationaloliveoil.org/
[3] Aparicio, R.; Ferreiro, L.; Alonso, V., Effect of climate on the chemical composition of virgin olive oil. *Analytica Chimica Acta* 1994, 292, (3), 235-241.
[4] Pardo, J. E.; Cuesta, M. A.; Alvarruiz, A., Evaluation of potential and real quality of virgin olive oil from the designation of origin "Aceite Campo de Montiel" (Ciudad Real, Spain). *Food Chemistry* 2007, 100, (3), 977-984.
[5] Marini, F.; Magrì, A. L.; Bucci, R.; Balestrieri, F.; Marini, D., Class-modeling techniques in the authentication of Italian oils from Sicily with a Protected Denomination of Origin (PDO). *Chemometrics and Intelligent Laboratory Systems* 2006, 80, (1), 140-149.
[6] Marini, F.; Magrì, A. L.; Bucci, R.; Magrì, A. D., Use of different artificial neural networks to resolve binary blends of monocultivar Italian olive oils. *Analytica Chimica Acta* 2007, 599, (2), 232-240.
[7] Ollivier, D.; Artaud, J.; Pinatel, C.; Durbec, J. P.; Guerere, M., Differentiation of French virgin olive oil RDOs by sensory characteristics, fatty acid and triacylglycerol compositions and chemometrics. *Food Chemistry* 2006, 97, (3), 382-393.
[8] Aranda, F.; Gomez-Alonso, S.; Rivera Del Alamo, R. M.; Salvador, M. D.; Fregapane, G., Triglyceride, total and 2-position fatty acid composition of Cornicabra virgin olive oil: Comparison with other Spanish cultivars. *Food Chemistry* 2004, 86, (4), 485-492.
[9] Alves, M. R.; Cunha, S. C.; Amaral, J. S.; Pereira, J. A.; Oliveira, M. B., Classification of PDO olive oils on the basis of their sterol

composition by multivariate analysis. *Analytica Chimica Acta* 2005, 549, (1-2), 166-178.
[10] Lopez Ortiz, C. M.; Prats Moya, M. S.; Berenguer Navarro, V., A rapid chromatographic method for simultaneous determination of β-sitosterol and tocopherol homologues in vegetable oils. *Journal of Food Composition and Analysis* 2006, 19, (2-3), 141-149.
[11] Cichelli, A.; Pertesana, G. P., High-performance liquid chromatographic analysis of chlorophylls, pheophytins and carotenoids in virgin olive oils: chemometric approach to variety classification. *Journal of Chromatography A* 2004, 1046, (1-2), 141-146.
[12] Aparicio, R.; Aparicio-Ruiz, R., Authentication of vegetable oils by chromatographic techniques. *Journal of Chromatography A* 2000, 881, (1-2), 93-104.
[13] Haddada, F. M.; Manai, H.; Daoud, D.; Fernandez, X.; Lizzani-Cuvelier, L.; Zarrouk, M., Profiles of volatile compounds from some monovarietal Tunisian virgin olive oils. Comparison with French PDO. *Food Chemistry* 2007, 103, (2), 467-476.
[14] Temime, S. B.; Campeol, E.; Cioni, P. L.; Daoud, D.; Zarrouk, M., Volatile compounds from Chétoui olive oil and variations induced by growing area. *Food Chemistry* 2006, 99, (2), 315-325.
[15] Lachenmeier, D. W.; Frank, W.; Humpfer, E.; Schafer, H.; Keller, S.; Mortter, M.; Spraul, M., Quality control of beer using high-resolution nuclear magnetic resonance spectroscopy and multivariate analysis. *European Food Research and Technology* 2005, 220, (2), 215-221.
[16] Mannina, L.; Segre, A., High resolution nuclear magnetic resonance: From chemical structure to food authenticity. *Grasas y Aceites* 2002, 53, (1), 22-33.
[17] Woodcock, T.; Downey, G.; O'Donnell, C. P., Near infrared spectral fingerprinting for confirmation of claimed PDO provenance of honey. *Food Chemistry* 2009, 114, (2), 742-746.
[18] Reid, L. M.; Woodcock, T.; O'Donnell, C. P.; Kelly, J. D.; Downey, G., Differentiation of apple juice samples on the basis of heat treatment and variety using chemometric analysis of MIR and NIR data. *Food Research International* 2005, 38, (10), 1109-1115.
[19] Baeten, V.; Pierna, J. A. F.; Dardenne, P.; Meurens, M.; Garcia-Gonzalez, D. L.; Aparicio-Ruiz, R., Detection of the presence of hazelnut oil in olive oil by FT-Raman and FT-MIR spectroscopy. *Journal of Agricultural and Food Chemistry* 2005, 53, (16), 6201-6206.

[20] Yang, H.; Irudayaraj, J.; Paradkar, M. M., Discriminant analysis of edible oils and fats by FTIR, FT-NIR and FT-Raman spectroscopy. *Food Chemistry* 2005, 93, (1), 25-32.
[21] Lopez-Diez, E. C.; Bianchi, G.; Goodacre, R., Rapid quantitative assessment of the adulteration of virgin olive oils with hazelnut oils using Raman spectroscopy and chemometrics. *Journal of Agricultural and Food Chemistry* 2003, 51, (21), 6145-6150.
[22] Yang, H.; Irudayaraj, J., Comparison of near-infrared, Fourier transform-infrared, and Fourier transform-Raman methods for determining olive pomace oil adulteration in extra virgin olive oil. *Journal of the American Oil Chemists' Society* 2001, 78, (9), 889-895.
[23] Vaclavik, L.; Cajka, T.; Hrbek, V.; Hajslova, J., Ambient mass spectrometry employing direct analysis in real time (DART) ion source for olive oil quality and authenticity assessment. *Analytica Chimica Acta* 2009, 645, (1-2), 56-63.
[24] Cajka, T.; Hajslova, J.; Pudil, F.; Riddellova, K., Traceability of honey origin based on volatiles pattern processing by artificial neural networks. *Journal of Chromatography A* 2009, 1216, (9), 1458-1462.
[25] Stanimirova, I.; Üstün, B.; Cajka, T.; Riddellova, K.; Hajslova, J.; Buydens, L. M. C.; Walczak, B., Tracing the geographical origin of honeys based on volatile compounds profiles assessment using pattern recognition techniques. *Food Chemistry* 2010, 118, 171-176.
[26] Vaz-Freire, L. T.; da Silva, M. D. R. G.; Freitas, A. M. C., Comprehensive two-dimensional gas chromatography for fingerprint pattern recognition in olive oils produced by two different techniques in Portuguese olive varieties Galega Vulgar, Cobrançosa e Carrasquenha. *Analytica Chimica Acta* 2009, 633, (2), 263-270.
[27] Martins-Lopes, P.; Gomes, S.; Santos, E.; Guedes-Pinto, H., DNA markers for Portuguese olive oil fingerprinting. *Journal of Agricultural and Food Chemistry* 2008, 56, (24), 11786-11791.
[28] Ranalli, A.; Contento, S.; Marchegiani, D.; Pardi, D.; Pardi, D.; Girardi, F., Effects of "genetic store" on the composition and typicality of extra-virgin olive oil: Traceability of new products. *Advances in Horticultural Science* 2008, 22, (2), 110-115.
[29] Reid, L. M.; O'Donnell, C. P.; Downey, G., Recent technological advances for the determination of food authenticity. *Trends in Food Science & Technology* 2006, 17, (7), 344-353.
[30] Galtier, O.; Dupuy, N.; Le Dréau, Y.; Ollivier, D.; Pinatel, C.; Kister, J.; Artaud, J., Geographic origins and compositions of virgin olive oils

determinated by chemometric analysis of NIR spectra. *Analytica Chimica Acta* 2007, 595, (1-2), 136-144.
[31] Petrakis, P. V.; Agiomyrgianaki, A.; Christophoridou, S.; Spyros, A.; Dais, P., Geographical characterization of Greek virgin olive oils (cv. Koroneiki) using ^{1}H and ^{31}P NMR fingerprinting with canonical discriminant analysis and classification binary trees. *Journal of Agricultural and Food Chemistry* 2008, 56, (9), 3200-3207.
[32] Rezzi, S.; Axelson, D. E.; Heberger, K.; Reniero, F.; Mariani, C.; Guillou, C., Classification of olive oils using high throughput flow ^{1}H-NMR fingerprinting with principal component analysis, linear discriminant analysis and probabilistic neural networks. *Analytica Chimica Acta* 2005, 552, (1-2), 13-24.
[33] Alonso-Salces, R. M.; Heberger, K.; Holland, M. V.; Moreno-Rojas, J. M.; Mariani, C.; Bellan, G.; Reniero, F.; Guillou, C., Multivariate analysis of NMR fingerprint of the unsaponifiable fraction of virgin olive oils for authentication purposes. *Food Chemistry* 2010, 118, 956-965.
[34] Christophoridou, S.; Dais, P.; Tseng, L. I. H.; Spraul, M., Separation and identification of phenolic compounds in olive oil by coupling high-performance Liquid Chromatography with Postcolumn Solid-Phase Extraction to Nuclear Magnetic Resonance Spectroscopy (LC-SPE-NMR). *Journal of Agricultural and Food Chemistry* 2005, 53, (12), 4667-4679.
[35] O'Leary, M. H., Environmental effects on carbon isotope fractionation in terrestrial plants. *Stable Isotopes in the Biosphere* 1995, 78-91.
[36] Ziegler, H.; Osmond, C. B.; Stichler, W.; Trimborn, P., Hydrogen isotope discrimination in higher plants: Correlations with photosynthetic pathway and environment. *Planta* 1976, 128, (1), 85-92.
[37] Royer, A.; Gerard, C.; Naulet, N.; Lees, M.; Martin, G. J., Stable isotope characterization of olive oils. I - Compositional and carbon-13 profiles of fatty acids. *Journal of the American Oil Chemists' Society* 1999, 76, (3), 357-363.
[38] Angerosa, F.; Breas, O.; Contento, S.; Guillou, C.; Reniero, F.; Sada, E., Application of stable isotope ratio analysis to the characterization of the geographical origin of olive oils. *Journal of Agricultural and Food Chemistry* 1999, 47, (3), 1013-1017.
[39] Aramendia, M. A.; Marinas, A.; Marinas, J. M.; Moreno, J. M.; Moalem, M.; Rallo, L.; Urbano, F. J., Oxygen-18 measurement of Andalusian olive oils by continuous flow pyrolysis/isotope ratio mass

spectrometry. *Rapid Communications in Mass Spectrometry* 2007, 21, (4), 487-496.
[40] Camin, F.; Larcher, R.; Perini, M.; Bontempo, L.; Bertoldi, D.; Gagliano, G.; Nicolini, G.; Versini, G., Characterisation of authentic Italian extra-virgin olive oils by stable isotope ratios of C, O and H and mineral composition. *Food Chemistry* 2010, 118, (4), 901-909.
[41] Camin, F.; Larcher, R.; Nicolini, G.; Bontempo, L.; Bertoldi, D.; Perini, M.; Schlicht, C.; Schellenberg, A.; Thomas, F.; Heinrich, K.; Voerkelius, S.; Horacek, M.; Ueckermann, H.; Froeschl, H.; Wimmer, B.; Heiss, G.; Baxter, M.; Rossmann, A.; Hoogewerff, J., Isotopic and elemental data for tracing the origin of European olive oils. *Journal of Agricultural and Food Chemistry* 2010, 58, (1), 570-577.
[42] Hoffman, R. E., Standardization of chemical shifts of TMS and solvent signals in NMR solvents. *Magnetic Resonance in Chemistry* 2006, 44, (6), 606-616.
[43] Berrueta, L. A.; Alonso-Salces, R. M.; Héberger, K., Supervised pattern recognition in food analysis. *Journal of Chromatography A* 2007, 1158, (1-2), 196-214.
[44] D'Imperio, M.; Mannina, L.; Capitani, D.; Bidet, O.; Rossi, E.; Bucarelli, F. M.; Quaglia, G. B.; Segre, A., NMR and statistical study of olive oils from Lazio: A geographical, ecological and agronomic characterization. *Food Chemistry* 2007, 105, (3), 1256-1267.
[45] Guillen, M. D.; Ruiz, A., High resolution H-1 nuclear magnetic resonance in the study of edible oils and fats. *Trends in Food Science & Technology* 2001, 12, (9), 328-338.
[46] Mannina, L.; Sobolev, A. P.; Segre, A., Olive oil as seen by NMR and chemometrics. *Spectroscopy Europe* 2003, 15, (3), 6-14.
[47] Sacchi, R.; Patumi, M.; Fontanazza, G.; Barone, P.; Fiordiponti, P.; Mannina, L.; Rossi, E.; Segre, A. L., A high-field H-1 nuclear magnetic resonance study of the minor components in virgin olive oils. *Journal of the American Oil Chemists' Society* 1996, 73, (6), 747-758.
[48] Schmidt, H. L.; Eisenreich, W., Systematic and regularities in the origin of 2H patterns in natural compounds. *Isotopes in environmental and health studies* 2001, 37, (3), 253-254.
[49] Schmidt, H. L.; Werner, R. A.; Eisenreich, W., Systematics of 2H patterns in natural compounds and its importance for the elucidation of biosynthetic pathways. *Phytochemistry Reviews* 2003, 2, (1-2), 61-85.
[50] DG-JRC Institute for the Protection and Security of the Citizen. MARS-project: Meteodata. http://www.marsop.info/marsop3.

[51] Breas, O.; Guillou, C.; Reniero, F.; Sada, E.; Angerosa, F., Oxygen-18 measurement by continuous flow pyrolysis/isotope ratio mass spectrometry of vegetable oils. *Rapid Communications in Mass Spectrometry* 1998, 12, (4), 188-192.

[52] Rossmann, A.; Schmidt, H. L.; Reniero, F.; Versini, G.; Moussa, I.; Merle, M. H., Stable carbon isotope content in ethanol of EC data bank wines from Italy, France and Germany. *Zeitschrift Fur Lebensmittel-Untersuchung Und-Forschung* 1996, 203, (3), 293-301.

[53] Owen, R. W.; Haubner, R.; Mier, W.; Giacosa, A.; Hull, W. E.; Spiegelhalder, B.; Bartsch, H., Isolation, structure elucidation and antioxidant potential of the major phenolic and flavonoid compounds in brined olive drupes. *Food and Chemical Toxicology* 2003, 41, (5), 703-717.

[54] Esbensen, K. H.; Guyot, D.; Westad, F.; Houmøller, L. P., *Multivariate data analysis in practice: an introduction to multivariate data analysis and experimental design.* 5th ed.; Camo Process AS: Oslo, Norway, 2002; p 598.

INDEX

A

accounting, 12
acid, 3, 53
acidity, 1
alcohols, 2, 11, 32
ANOVA, ix, 7, 26
antioxidant, 58
assessment, 55
authentication, 3, 4, 12, 27, 29, 35, 45, 53, 56
authenticity, 2, 3, 4, 26, 44, 54, 55
authorities, 2, 4, 26
authors, 11, 16, 22, 51

B

beer, 54
binary blends, 53
biosynthetic pathways, 20, 57
blends, 53

C

calibration, 7
capillary, 6
carbon, 7, 56, 58
carotenoids, 2, 54
case study, 26, 42
chemometrics, 53, 55, 57
chloroform, 5, 6
chromatographic technique, 3, 54
chromatography, ix, 55
class, vii, 3, 31, 32, 42
classification, 4, 8, 11, 12, 26, 31, 32, 34, 39, 42, 43, 44, 54, 56
climate, 1, 3, 16, 20, 22, 33, 53
climatic factors, 20
clusters, 12, 22
CO_2, 7, 20
color, iv
community, 2
composition, 2, 3, 20, 33, 42, 53, 54, 55, 57
compounds, 2, 3, 11, 32, 41, 43, 49, 54, 55, 56, 57, 58
copyright, iv
cost, 42
country of origin, 2, 44
covering, 23
cultivation, 33
Cyprus, 4, 5, 11

D

damages, iv
DART, 55
data analysis, 6, 7, 26, 58

data set, 7, 8, 12, 26, 27, 29, 31, 32, 33, 34, 44, 47
datasets, 26
detection, 4, 42
deviation, 7
discriminant analysis, vii, ix, 4, 7, 28, 56
discrimination, 3, 4, 12, 33, 42, 44, 56
diversity, 2
DNA, 3, 55

E

elucidation, 57, 58
environmental conditions, 3, 20, 42
environmental factors, 22
ethanol, 58
European Commission, 2, 51
European Union, 1, 2
experimental design, 58
external validation, vii, 8, 31, 34, 43, 44
extraction, ix, 2, 3, 26

F

farmers, 16
fatty acids, 3, 11, 38, 56
fingerprints, vii, 4
flavor, 1
food products, 12
France, vii, 2, 4, 5, 11, 16, 17, 51, 58
fraud, 2, 42
FTIR, 55
funding, 51

G

Germany, 5, 6, 58
glycerol, 6, 32, 38, 43
Greece, vii, 1, 4, 5, 11, 16, 17, 42, 43, 44, 45, 46, 47, 48
grouping, 22

H

harvesting, 16, 22
heat treatment, 54
hydrocarbons, 11
hydrogen, 7, 20

I

imbalances, 31, 32
induction, ix, 6
interdependence, 2
international standards, 7
isolation, 3
isotope, vii, 3, 7, 16, 56, 57, 58
issues, 2
Italy, vii, 1, 4, 5, 6, 11, 16, 17, 18, 20, 22, 23, 26, 33, 42, 43, 44, 45, 46, 47, 48, 51, 58

L

labeling, 2, 26, 42
liquid chromatography, ix

M

magnetic resonance, ix, 54, 57
magnetic resonance spectroscopy, 54
markers, 55
mass spectrometry, vii, 3, 55, 57, 58
matrix, 8, 12, 33
median, 16, 18, 21, 22, 23, 44
Mediterranean, vii, 1, 5, 11, 12
metabolism, 3
microclimate, vii, 16
modeling, vii, 3, 8, 12, 34, 44, 53
moisture, 4
multivariate data analysis, 6, 26, 58

N

neural network, 53, 55, 56
neural networks, 53, 55, 56
NIR, ix, 3, 54, 55, 56
NIR spectra, 3, 56
NMR, vii, ix, 3, 4, 5, 6, 7, 11, 12, 14, 26, 27, 29, 32, 35, 36, 37, 39, 42, 43, 44, 45, 46, 47, 48, 56, 57
Norway, 8, 58
nuclear magnetic resonance, ix, 54, 57

O

oil, 1, 2, 3, 4, 5, 6, 7, 11, 18, 26, 42, 43, 44, 51, 53, 54, 55, 56, 57
oil samples, 51
olive oil, vii, ix, 1, 3, 4, 5, 6, 7, 11, 16, 18, 26, 33, 42, 43, 44, 51, 53, 54, 55, 56, 57
optimization, 8
organic solvents, 3
overlap, 11
oxidation, 6

P

parallel, 23
parameter, 8
Pareto, 8, 26, 28, 30, 31, 32
partial least-squares, ix, 7
pathways, 20, 57
pattern recognition, 4, 6, 26, 27, 29, 31, 32, 35, 45, 55, 57
PCA, ix, 4, 7, 12, 13, 34
performance, 54, 56
permission, iv
phytosterols, 2
plants, 16, 56
PLS, vii, ix, 4, 7, 8, 26, 27, 28, 29, 30, 31, 32, 33, 34, 35, 36, 37, 38, 39, 43, 44, 45, 46, 47, 48
Portugal, 2

prevention, 4
principal component analysis, vii, ix, 4, 7, 56
probability, 28, 31
probe, 6
producers, 5
project, 4, 5, 51, 57
properties, 1, 53
protons, 11, 32, 38, 43
pyrolysis, 56, 58

R

rainfall, vii, 16, 18, 20, 22, 24
Raman spectroscopy, 55
ratio analysis, 56
raw materials, 1
reagents, 3
real time, 55
recognition, vii, 4, 6, 8, 26, 27, 29, 31, 32, 35, 55, 57
recommendations, iv
recycling, 6
regression, 32, 38, 42, 43, 44
regulatory bodies, 4, 26
reliability, 8
requirements, 2, 26
resolution, 54, 57
rights, iv
root-mean-square, ix

S

saturated fat, 38
saturated fatty acids, 38
scaling, 26, 28, 30, 32
scarcity, 22
security forces, 2
SIC, 18, 19
signals, 11, 14, 32, 38, 43, 57
silver, 7
software, 7
solvents, 3, 57

Index

space, 13, 26
Spain, vii, 1, 4, 5, 11, 16, 17, 22, 42, 43, 44, 45, 46, 47, 48, 53
species, 2
spectroscopy, 54, 55
stable isotopes, 42
standard deviation, 7
sterols, 3, 11
storage, 2
strategy, 8
substitution, 11
Sweden, 8
Syria, 4, 5, 11

T

temperature, vii, 4, 6, 16, 20, 24

terpenes, 11, 32
territory, 23
tin, 7
tocopherols, 11
training, 8, 31, 33, 34, 43, 44
transformation, ix, 6
treatment methods, 3
triglycerides, 2, 3, 11, 14, 15, 38, 43, 48
Turkey, 4, 5, 11

V

validation, vii, 9, 31, 34, 43, 44
valleys, 24
variations, 54
vector, 26
vegetable oil, 54, 58